Marco Mernberger

Graph-based approaches to protein structure comparison

Marco Mernberger

Graph-based approaches to protein structure comparison

From local to global similarity

Südwestdeutscher Verlag für Hochschulschriften

Impressum / Imprint
Bibliografische Information der Deutschen Nationalbibliothek: Die Deutsche Nationalbibliothek verzeichnet diese Publikation in der Deutschen Nationalbibliografie; detaillierte bibliografische Daten sind im Internet über http://dnb.d-nb.de abrufbar.

Bibliographic information published by the Deutsche Nationalbibliothek: The Deutsche Nationalbibliothek lists this publication in the Deutsche Nationalbibliografie; detailed bibliographic data are available in the Internet at http://dnb.d-nb.de.

Coverbild / Cover image: www.ingimage.com

Verlag / Publisher:
Südwestdeutscher Verlag für Hochschulschriften
ist ein Imprint der / is a trademark of
AV Akademikerverlag GmbH & Co. KG
Heinrich-Böcking-Str. 6-8, 66121 Saarbrücken, Deutschland / Germany
Email: info@svh-verlag.de

Herstellung: siehe letzte Seite /
Printed at: see last page
ISBN: 978-3-8381-3452-9

Zugl. / Approved by: Marburg, Uni, Diss. 2011

"The Road goes ever on and on
Down from the door where it began.
Now far ahead the Road has gone,
And I must follow, if I can,
Pursuing it with eager feet,
Until it joins some larger way
Where many paths and errands meet.
And whither then? I cannot say."

J.R.R. Tolkien, *The Lord Of The Rings*

Contents

List of Figures

LIST OF FIGURES

List of Tables

LIST OF TABLES

1

Introduction

The advent of more and more advanced sequencing technologies has brought a vast amount of information detailing genes in a variety of different organisms. With each year, this data is expanding rapidly, producing novel sequences for which no functional annotation exists.

While the acquisition of protein structure data is far behind that rate, the Protein Data Bank (PDB) (Berman et al., 2000) is nevertheless growing exponentially with each passing year (Klebe, 2009), thanks to advances in the field of nuclear magnetic resonance spectroscopy (Pellecchia et al., 2002) and high-throughput crystallography (Blundell et al., 2002). Several structural genomics projects across the globe aim at closing the gap between sequence and structure knowledge by experimentally determining the structure of a large number of proteins by high-throughput approaches as fast and as accurate as possible. While all of these projects have their own agenda, this ultimately serves to increase the coverage of protein structure space significantly, yielding a large number of structures for which no functional annotation is available (Chandonia and Brenner, 2006; David et al., 2011).

This is in contrast to the classical approach to protein structure analysis which typically involves starting with a protein of interest, collecting functional information by conducting biochemical experiments and then turn to the protein structure to rationalize the function (Thornton et al., 2000). As a result, the need for robust automated prediction methods that are capable of deriving a prediction of function for unknown proteins is as great as ever. Moreover, the possibility to draw upon structural information to infer protein function for proteins without functional annotation is becoming increasingly viable, thanks to the continuing growth of the PDB.

1. INTRODUCTION

The steady improvement of structure prediction tools (Ben-David et al., 2009; Qian et al., 2007; Wang et al., 2010) based on sequence information offers another promising source for structure information. For many cases, sophisticated modeling approaches can already generate very accurate structure predictions, although the *de novo* prediction of structures is still an open problem (Kryshtafovych et al., 2009).

Given this increase in available data, it is hardly surprising that structural bioinformatics has gained increasing attention in the past years (Andreeva and Murzin, 2010; Berman et al., 2000, 2009; Pérot et al., 2010), the major field of application being the inference and analysis of protein function. The inference of function of unknown proteins plays a central role in life sciences in general and pharmaceutical chemistry in particular. In this regard, the comparison of proteins is a central task.

Generally speaking, prediction of protein function is either done on the sequence level or the structure level. Sequence-based comparison is usually the first method of choice, owing to the observation that proteins with an amino acid sequence similarity above 40 % tend to have similar functions (Todd et al., 2001). For this task, a large variety of different algorithmic approaches are available and widely used (Altschul et al., 1997; Edgar, 2004; Hannenhalli and Russell, 2000; Jensen et al., 2003; Larkin et al., 2007; Notredame et al., 2000; Pearson, 1991; Sjölander, 2004).

Below this threshold however, protein function is much less conserved (Whisstock and Lesk, 2004). As a result, prediction accuracy declines for proteins whose sequence identity falls below a certain percentage (Lee et al., 2007; Rost, 2002; Tian and Skolnick, 2003). Where sequence-based methods fail to provide a functional prediction, for example in case of *orphan proteins*, structure-based approaches can allow us to gain further insights.

Structural similarity can still exist even if the corresponding sequences show low similarity. This is not as surprising as one might think. While it is true that protein structure is determined by the amino acid sequence, it has been shown that only a small fraction of amino acids are crucial to stabilizing a certain three-dimensional fold (Guo et al., 2004; Russ et al., 2005). One also has to mention that the number of viable protein folds is much smaller as sequence variability would suggest. Some low estimates assume merely 1,000 (Leonov et al., 2003; Wang, 1998) folds, while other, more realistic numbers range between 4,000 to 10,000 (Govindarajan et al., 1999; Grant et al., 2004; Liu et al., 2004) folds. Thus, it is save to say that a high sequence similarity indicates functional similarity, but the opposite is not necessarily true.

In the field of pharmaceutical chemistry, the structural analysis of proteins is also of special interest. Beyond the inference of protein function, one is typically interested in a more detailed analysis of the protein active site in order to gain a better understanding of the mechanisms governing enzyme activity and the interaction between protein and ligand or substrate.

As sequence-based approaches can hardly be used to pinpoint the spatial positioning of functionally important residues, the use of structure-based approaches might again provide further insights into these problems. Moreover, subtle differences in the structural composition of active sites can cause differences in enzyme activity, affinity to a certain substrate or ligand and even alter the catalyzed biochemical reaction (Jost et al., 2010; Zou et al., 2010). Thus, it is questionable whether sequence-based approaches alone are always sufficient to gain an understanding of these mechanisms, as the protein structure carries potentially much more direct information than sequence alone.

1.1 Aspects of protein structure comparison

A major task in structural bioinformatics is the comparison of protein structures. While protein structure data is much less abundant than protein sequence information, the comparison of structures has nevertheless been attempted with great vigor and effort in the structural bioinformatics community. Unsurprisingly, a huge number of different algorithms have been proposed during the last 15 years that employ fundamentally different principles.

These approaches can be divided into two major categories, much in analogy to sequence comparison methods. In sequence comparison, DNA and amino acid sequences are mostly compared in terms of global alignments (Needleman and Wunsch, 1970), which align every amino acid or nucleotide in the sequences of interest, or local alignments (Smith and Waterman, 1981), whose goal is to find and align similar common subsequences. The former strategy is most useful for evolutionary related sequences of roughly equal length, while the latter approach is usually chosen for sequences that are dissimilar but suspected to contain similar subsequences, corresponding to common motifs or functional domains. Hybrid methods have also been proposed (Brudno et al., 2003).

This distinction is also valid for protein structure comparison. One group of approaches generally aims at comparing complete protein structures, which is usually referred to as *global structure comparison*. Other approaches specifically focus on certain regions of interest, for example protein active sites, catalytic triads, protein-protein interfaces or protein binding sites. This latter group of approaches is consequently described as *local structure comparison* (Watson et al., 2009).

Global structure comparison methods compare the complete tertiary structure of proteins in terms of their fold, using geometric approaches as well as algorithms based on derived representations, such as secondary structure elements. An overview of these approaches will be given in Chapter 2. The rationale behind these approaches is the assumption, that the so-called *fold space*, the set of all existent tertiary folds (Scheeff and Fink, 2009), is much smaller than the diversity of sequences implies (Govindarajan et al., 1999; Grant et al., 2004; Leonov et al., 2003; Liu et al., 2004; Wang, 1998).

Local structure comparison instead focuses on functionally relevant substructures of the complete proteins. Among these methods are so-called template-based and surface-based approaches, which will be discussed in the next chapter. The motivation here is the assumption, that the overall fold is less important for a specific function, than the local region, where the catalytic center is located. Indeed, examples are known in which similar folds catalyze different biochemical reactions (Copley et al., 2004; Nagano et al., 2002; Orengo et al., 1999).

Both principles have their merits. While global structure comparison can in many cases successfully identify functional similarities and uncover important functional domains (Thornton, 2001), they mostly consider a much lower level of detail, as usually only protein backbones are compared. Local structure comparison instead trades the overall picture for a more detailed view of the substructures, as usually side chain positions are considered as well (Watson et al., 2009).

In pharmaceutical chemistry, especially local structure comparison is of interest, as a major goal is the design of usually small molecules as drug candidates that are able to interact and modulate the function of certain target proteins. To this end, a more detailed representation of the targets is obviously necessary.

In this work, the focus will be on the comparison of protein binding sites, a special case of local protein structure comparison, hence the approaches and experiments presented in this thesis can be considered as local protein comparison approaches, although the developed algorithms are not limited to this kind of application and might be interesting for other tasks as well.

1.2 Protein structure comparison in pharmaceutical chemistry

In pharmaceutical chemistry, the analysis and comparison of structures is a useful methodology in knowledge-based drug design, especially in the context of computer-assisted drug design (CADD). CADD offers a great arsenal of methods that helps to design new drug candidates and assess their potential activity. Combinatorial chemistry allows for the computerized design of potential agents by combining libraries of chemical groups (Corbett et al., 2006). Molecular dynamics deals with the prediction of conformations and conformational changes of molecules using statistical simulations (Karplus and McCammon, 2002). Knowledge-based drug design utilizes structure information of ligands and proteins.

Roughly speaking, knowledge-based drug design can be divided in two major categories: Ligand-based drug design and receptor-based drug design (Schneider and Fechner, 2005). Ligand-based drug design exploits knowledge of existing ligand structures that are known to bind to a protein of interest in the development of new drug candidates. Methods in this category include for example the construction of pharmacophore models (Langer and Hoffmann, 2006) or quantitative structure-activity relationships (QSAR) (Dudek et al., 2006). Receptor-based drug design instead utilizes knowledge of the target structure. This involves database searches and virtual screening (to find potential ligands for a given target), docking techniques (to predict the binding affinity of a specific drug candidate to its target) (Morris et al., 1998) or structure-based *de novo* ligand design (Böhm, 1992; Zhu et al., 2001). The comparison of protein structures and protein binding sites also belongs into this category and is useful for several different tasks.

One primary objective in modern drug development is the identification of new *druggable* targets, that are related to a certain disease. In this context, the term *druggable* refers to targets whose function can be modulated by small chemical compounds that can be used as therapeutic agents. In most cases, these biological targets are proteins, such as enzymes, ion channels, hormone receptors, transport proteins and others, but also nucleic acids can be targets (Chen et al., 2002; Zhu et al., 2010). The characterization and prediction of protein function is an important aspect of the search of potential new drug targets, especially, if one is interested in a high selectivity towards pathogens, in which case one would typically select targets that are characteristic of the pathogen. This can be attempted using either sequence or structure comparison.

1. INTRODUCTION

The ultimate goal in knowledge-based drug design is the development of compounds that are likely to interfere with a given target. Typically, these molecules act as ligands that bind to the target structures and modulate their function, i.e. by acting as agonists, inverse agonists or antagonists for receptors, or as enzyme inhibitors and allosteric effectors (Klebe, 2009). In the case of protein targets, these interactions between ligand and protein typically occur in clefts on the protein surface referred to as protein binding sites (Laskowski et al., 1996; Peters et al., 1996). The binding of a certain ligand in a protein binding site occurs in a very specific way, which is often described by a key-lock analogy, a metaphor that traces back to the German chemist and Nobel laureate Emil Fischer in 1894. More precisely, the ligand possesses a suitable size and form to fit optimally in the spatial confinement of the binding site and exhibits complementary chemical properties to the surface properties of the binding site.

Ideally, a novel candidate should be "druglike", i.e. exhibit characteristics such as high efficacy, potency and bioavailability, minimal side effects and a high metabolic stability. Different rules and scoring functions exist to appraise the druglikeness of a compound, such as the Lipinski rule-of-five (Lipinski et al., 1997) or the lipophilic efficiency (Ryckmans et al., 2009). Other desirable properties are a high specificity and affinity to the target structure as well as a high selectivity, though also highly promiscuous therapeutic agents exist that interfere with multiple targets, so-called "dirty drugs" (Klebe, 2009). Especially for these latter properties, the analysis of protein binding sites and ligand-protein interactions plays an important role, as such analyses are vital for the understanding of the governing principles that are responsible for the addressability of a target. For this reason, many surface-based databases of protein binding sites have been developed (Binkowski et al., 2003b; Kinoshita and Nakamura, 2003; Schmitt et al., 2002; Shulman-Peleg et al., 2004). The analysis and comparison of such binding sites is instrumental for understanding the chemical basis of the protein-ligand interaction and the mode of action that determines the protein function.

In recent years, the characterization of protein families has become more and more important in this regard, based on the assumption that related proteins bind similar ligands (Naumann and Matter, 2002). Such so-called *chemogenomics* aim at the identification of the structural and physicochemical properties of a protein that influence selectivity and specificity by ascertaining the commonalities and differences of related proteins (Bredel and Jacoby, 2004; Mestres, 2004). A comparison of protein binding sites in this context is a reasonable strategy, since it focuses on the region of interest instead of comparing complete proteins on sequence or fold level.

Another important aspect in knowledge-based drug design is the identification and prediction of cross-reactivities. In a study of 2006, the cost of bringing a new therapeutic agent to market was estimated around 500 million to two billion US dollar, depending on developing company and therapy (Adams and Brantner, 2006). The prediction of proteins that are likely to interact with a certain new drug candidate can help to identify potential cross reactivities long before expensive experimental studies are conducted, thus helping to filter out potentially harmful agents and lowering production cost.

Finally, it should be stated that the comparison of molecular structures is not only of interest for proteins. For example, in ligand-based drug design, a comparison of ligand structures can be used to ascertain the minimal structural characteristics a drug candidate must exhibit in order to interact with the target. This information is typically used to built a pharmacophore model which in turn is used to design new candidates fulfilling the necessary requirements (Langer and Hoffmann, 2006).

1.3 Graph theory and structure comparison

This thesis focuses on the comparative analysis of protein binding sites as a special case of molecular structure data. This obviously raises the question, how molecular structure data should be modeled in order to make it amenable to algorithmic methods. In the realm of structural bioinformatics, many modeling concepts have been used, the most common among them being secondary structure elements, direct geometrical data, distance matrices and graphs (Marti-Renom et al., 2009; Watson et al., 2009).

As will be outlined in Chapter 2, graphs represent a versatile and powerful framework for the modeling of structured data. Graphs have been widely used in chemoinformatics for the modeling of chemical compounds (Balaban et al., 1976; Bunke and Jiang, 2000) as well as for the modeling of protein structure data in bioinformatics (Artymiuk et al., 1994; Kinoshita and Nakamura, 2003).

In the context of this work, protein binding sites will be modeled using undirected edge-weighted and node-labeled graphs. Thus, the comparison of protein binding sites translates to a comparison of graphs and the algorithms presented in this thesis basically constitute conceptually different approaches to graph comparison.

From a machine learning point of view, real-world objects are most commonly described by a set of attributes or features. Hardly surprising, many algorithms have been developed that build on this type of representation (Bishop, 2006). In the case of structured data, this can become problematic. While it is always possible to devise a function that maps structured objects to a set of structural descriptors, this usually incurs a loss of information, i.e. the global structure is lost and cannot be recovered from the descriptors. While this might not be a problem per se, it can lead to suboptimal results in cases where the overall structure itself is important. In this regard, graphs are more convenient for this work, since this data structure allows to model the dependencies between the constituents of an object without decomposing the global structure.

This is even more important, since the derivation of meaningful attributes is a problem of its own. Especially for protein binding sites, it is not possible yet to determine the functionally important regions of a binding site automatically in advance, based solely on protein structure information.

Graphs are used in a much wider context than the representation of molecular structures. For example, graphs play a major role as modeling concepts for biological networks, such as interaction networks (Berg and Lässig, 2004) or metabolomic networks (Kanehisa et al., 2004). Moreover, they are used beyond the realm of life science, for example to model social networks (Wasserman and Faust, 1994), HTML documents (Page et al., 1998) or the Internet itself (Borgwardt, 2007).

From this point of view, focusing on graph-based methods is interesting as well, as the approaches developed during this work could in principle be extended to other graph-based problems as well and the general ideas behind the approaches might be of interest for other fields of applications.

Particularly in the field of computer vision and pattern recognition, the presented methods might be of some interest, since the problem of comparing structured objects is commonly encountered and addressed in this field. A plethora of different approaches have been developed in this field as well, some also utilizing graph-based models (Conte et al., 2004).

1.4 Goals

The main goal of this thesis is the development and validation of new algorithms for the comparative analysis of protein binding sites or binding pockets. As outlined above, protein binding pockets play an important role in pharmaceutical chemistry. But the comparison of protein binding sites carries also some benefits for structural bioinformatics in general. By comparing protein binding sites rather than complete protein structures or sequences, one arguably focuses on the essential part of the structure that is responsible for its function. This reasoning is based on the assumption that the spatial composition of the binding site determines the capability of a protein to bind and interact with its substrates and ligands. Hence, important catalytic residues are oriented towards such a cavity.

In principle, the comparison of these binding pockets can be used to approach several different tasks, both in structural biology and knowledge-based drug design:

- Prediction of protein function: The biological function of a protein is largely influenced by the spatial structure of the active center and the architecture of their substrate binding site. Thus, retrieving similar binding pockets could in principle be used to predict the function of unknown proteins, in cases where sequence similarity or the overall fold might not yield a conclusive result (e.g. orphan structures).

- Prediction of cross-reactivities: Given a new drug candidate that binds to a certain target structure, it is important to identify potential cross-reactivities early in the developmental stage. By searching for similar binding sites in a database of cavities, potential cross-reactivities can be identified before expensive experimental studies are conducted.

- Identifying new target structures: When searching for new potential drug targets in a specific pathogen, a comparative analysis of protein binding sites could reveal structures that are sufficiently dissimilar from their human counterpart to allow the development of highly selective drug candidates.

- Uncovering distant evolution: The comparison of protein binding sites could in principle be used to uncover more distantly related proteins in the so-called "twilight zone" of proteins, where sequence similarity is too low to infer a common ancestor reliably. This, however, must be viewed critically, as it carries the risk of confusing a real hereditary relationship with convergent evolution.

- Characterizing protein families: Comparing protein binding sites can help to characterize protein families on a functional level in the context of chemogenomics.

As mentioned previously, protein binding sites will be modeled as graphs, more precisely edge-weighted and node-labeled graphs, building on the binding site representation used in CavBase (Kuhn et al., 2006; Schmitt et al., 2002), a database developed for the of storage and analysis of putative binding pockets. These are extracted by the LigSite algorithm (Hendlich et al., 1997) as clefts on the protein surface in experimentally determined protein structures derived from the PDB. This is illustrated in Fig. 1.1.

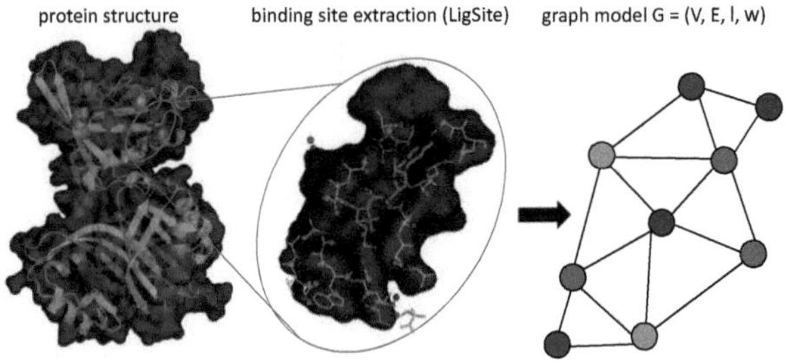

Figure 1.1: The modeling concept used in this work for the representation of protein binding sites. Binding sites are extracted from PDB structures using the LigSite algorithm and subsequently modeled as node-labeled and edge-weighted graphs.

In a previous work, such graph models have already been used to compare binding pockets using a greedy heuristic (Weskamp, 2007). One motivation for this work was to expand upon this representation and to improve the previously introduced approach. Thus, the comparison of protein binding sites translates to a comparison of graphs.

As outlined above, the comparison of DNA or amino acid sequence as well as the comparison of protein structure share some analogy, as both tasks are approached simultaneously on a local or a global scale. The comparison of protein binding pockets in the sense of protein structure comparison is regarded as a local comparison problem. But by regarding the protein binding sites as the actual entities to be compared, the dualism of global and local comparison can be carried one step further.

So far, binding sites as such have largely been compared using global strategies, i.e. by regarding complete binding sites as a whole. The heuristic approach by Weskamp (2007) is no exception, as it derives a global correspondence between binding sites in the form of a structural alignment, by comparing the graphs corresponding to these binding sites as a whole. But is this the only reasonable way to approach this problem?

In analogy to sequence comparison as well as protein structure comparison, one can alternatively approach the task of comparing protein binding sites by using a local graph comparison method, thus deriving a measure of similarity for binding pockets by comparing local properties of the corresponding graph models.

At this point, to avoid any terminological confusion, it should be stated that in this context, the terms *local* and *global* refer to the way in which the corresponding graphs are compared, regardless of whether a graph represents a whole molecular structure or only a part thereof. In other words, these terms are used in a purely methodological sense. This is in contrast to the way these terms are usually perceived in the field of structural bioinformatics, where a global comparison typically refers to comparing complete molecular structures and a local comparison to a comparison of substructures, such as domains of active sites.

Global approaches aim at detecting the largest common substructure of two or more binding sites sharing the same function. The rationale behind this is the notion that this must be the essential part for the interaction with the substrate or ligand. If this is the case, commonalities would be accurately detected by performing a global comparison.

On the other hand, a local strategy might be more useful, at least in some cases. Ligands in general are flexible molecules that might be bound in different conformations. Thus it is imaginable that parts of the ligand are addressed by different *subpockets* that might not necessarily be arranged in the same constellation. Moreover, determining the borders of a cavity is difficult and the functionally relevant part of a binding site might be just a fraction of the extracted cavity, if the cavity corresponds to a binding site at all.

Also, protein structures themselves are flexible and subjected to conformational changes and even mutations that do not necessarily cause a loss of function. Given that also structural data itself is noisy, due to measurement errors and resolution issues, it is obvious, that some tolerance towards structural variation is necessary. While local approaches will be less affected by such variation, it might be possible to combine this robustness with the benefits of a global algorithm, yielding a semi-global algorithm.

All these ideas have some merit and it remains to be tested, which kind of approach would have the greater benefit: Is a global method the only viable strategy or is a local approach perhaps more suitable for the task? How would a combination of both principles, a semi-global (or semi-local) method perform?

In order to shed some light on these questions, one purpose of this thesis is to develop different algorithms based on these three principles, i.e., global, local and semi-global methods, and to validate and analyze their performance in a comparative study to convey an idea of the benefits and problems associated with these principles.

1. INTRODUCTION

2

Related Work

As outlined above, the comparative analysis of proteins is a central task in life science in general. Hence it is not surprising, that a plethora of algorithms exist, that aim at a comparison of proteins on one level or another. Moreover, approaches in this field can either be viewed from an application point of view, focusing on the different aspects of protein comparison, or a methodological one, assessing the different algorithmic strategies involved.

The interdisciplinary nature of this study necessarily demands a consideration of both aspects and thus touches a variety of different research areas, from the field of pharmaceutical chemistry to structural genomics and computational biology. A discussion of relevant methods will necessarily remain incomplete, given the huge amount of activity in these fields. However, in the following, the most important methods are discussed briefly to place this thesis in the wider view of structure comparison.

2.1 Sequence-based approaches to protein comparison

Generally speaking, protein comparison is either done on the sequence level or the structure level. Sequence-based comparisons are widely used due to the nearly endless availability of sequence information (UniProt-Consortium, 2009). Significant similarities in sequence can usually provide a strong indication of functional and structural similarity, especially for highly similar sequences (Chothia and Lesk, 1986; Hark Gan et al., 2002; Wilson et al., 2000; Wood and Pearson, 1999).

Sequence comparison typically boils down to calculating sequence alignments. Among the first and most basic algorithms for pairwise sequence alignments are the Needleman-Wunsch (Needleman and Wunsch, 1970) and the Smith-Watermann (Smith and Waterman, 1981) algorithm. Both algorithms generate exact alignments with respect to a certain scoring function using dynamic programming. While the former generates a globally optimal alignment, that is, a global correspondence of the complete sequences, the latter calculates local alignments and thus identifies common subsequences in the amino acid composition.

For inference of function, these algorithms are of limited use due to their relatively high complexity. Given the large amount of available data, sequence alignments are more often approached by heuristics. Among the most widely used approaches are the well-known BLAST (Altschul et al., 1990), Psi-BLAST (Altschul et al., 1997) and FASTA (Pearson, 1991). While these methods cannot guarantee to find an optimal solution, they are typically very fast and provide good results.

For lower sequence identity it becomes difficult to detect relationships among proteins. More advanced pairwise methods that utilize sequence profiles (Gribskov et al., 1987), Hidden Markov Models (Eddy, 1996; Krogh and Brown, 1994) or a combination of both (Eddy, 1998) are more sensitive for lower sequence identities (Watson et al., 2009).

The use of multiple alignment techniques also helps to alleviate this problem to a certain degree (Park et al., 1998) and further improves the inference of function from sequence, as a multiple alignment also offers the possibility of detecting conserved and thus highly important residues in homologue proteins. The most prominent approaches also use HMM profiles, like ClustalW2 (Larkin et al., 2007). T-Coffee (Notredame et al., 2000), MAFFT (Katoh et al., 2002) or MUSCLE (Edgar, 2004) expand pairwise to multiple alignments by using the *progressive approach*, basically a greedy strategy following some sort of guiding information (e.g. phylogenetic distance, kmer distance).

Still, sequence comparison cannot be used to analyze the spatial location of functionally important residues, which is of great interest when assessing the biochemical function of a protein. Moreover, sequence alignments cannot always uncover the function of a protein correctly. While it has been shown that proteins with a sequence identity above 40 % tend to share similar function (Todd et al., 2001), the results of sequence comparison become more and more uncertain below that threshold (Lee et al., 2007; Rost, 2002; Tian and Skolnick, 2003; Whisstock and Lesk, 2004).

As was mentioned in the previous chapter, structural similarity can still exist while the sequence similarity is low, since only a small fraction of amino acids actually stabilize a certain fold and the greater part of the sequence is more or less irrelevant for the tertiary structure (Guo et al., 2004; Russ et al., 2005). Given that the number of viable protein folds is much smaller than sequence variation would suggest (Govindarajan et al., 1999; Grant et al., 2004; Leonov et al., 2003; Liu et al., 2004; Wang, 1998), structure comparison can provide further insights and uncover more remote similarities in these cases (Thornton, 2001; Zarembinski et al., 1998). Some advanced HMM profile-based approaches use structure information as well, e.g., to define protein families and detect homologues (Gene3D (Yeats et al., 2006)).

2.2 Protein structure comparison

In analogy to sequence comparison, approaches to protein structure comparison can roughly be divided into local and global methods. Global methods typically aim at deriving a structural correspondence between structures as a whole, while local methods usually focus on the comparison of functionally relevant parts of the molecules. Such a comparison usually involves the calculation of structural alignments, similar to sequence alignments.

Among the global methods are fold-based and geometric approaches, while local methods typically refers to template-based and surface-based approaches. As each group focuses on a different aspect of similarity, different strategies have been developed for the comparison and alignment of molecular data.

2.2.1 Fold-based and geometric structure comparison

Fold-based approaches typically compare the overall structure of proteins in terms of fold geometry and can thus be considered global approaches. The function of unknown proteins is inferred from the closest match in fold databases like CATH (Orengo et al., 1997), SCOP (Murzin et al., 1995), FSSP (Holm and Sander, 1996) or SUPERFAMILY (Gough and Chothia, 2002).

These approaches focus on different levels of detail and calculate different types of alignments. Geometric approaches represent protein structures on the level of atom coordinates and typically produce alignments of C_α atoms of residues, whereas other approaches use a higher level of abstraction, e.g. secondary structure elements (SSE). As the calculation of optimal alignments for complete protein structures is NP-complete (Lathrop, 1994), this task is usually tackled by heuristics. In the following, some of these approaches are discussed briefly.

Most prominent among the fold-based approaches is the FSSP-associated DALI method, which essentially uses contact maps of inter-residue distances derived from corresponding C_α-atoms. Alignments of structurally equivalent residues are generated from so-called contact patterns, equivalent regions in the matrices, by means of a Monte Carlo-approach (Holm and Park, 2000; Holm and Sander, 1995). SSAP compares inter-atomic distance vectors to derive a similarity measure for residues and calculates alignments using double dynamic programming (Orengo and Taylor, 1996; Taylor et al., 1994). Another algorithm called Structal instead focuses on pairwise residue distances and utilizes iterative dynamic programming (Gerstein and Levitt, 1996, 1998).

Among the methods build on SSE representations are VAST (Gibrat et al., 1996; Madej et al., 1995) which calculates vector alignments of secondary structures, MSD-Fold/SSM (Krissinel and Henrick, 2004a) and CATHEDRAL (Redfern et al., 2007). All three methods rely on graph theory to find a correspondence between folds by modeling the secondary structure in terms of graphs (SSEs are represented as graph nodes) and performing graph comparison. CATHEDRAL includes the GRATH algorithm as a component, while VAST uses PROTEP (Grindley et al., 1993), both of which are based on clique-enumeration approach (Bron and Kerbosch, 1973), while SSM utilizes its own subgraph isomorphism algorithm (Krissinel and Henrick, 2004b).

The MASS algorithm (Dror et al., 2003) calculates multiple alignments based on SSEs utilizing geometric hashing (Nussinov and Wolfson, 1991). The more recent GANGSTA+ (Guerler and Knapp, 2008) approach uses a combinatorial algorithm based on SSE contact maps to generate non-sequential secondary structure alignments while its predecessor employed a genetic algorithm (Kolbeck et al., 2006).

The combinatorial extension approach (CE) (Shindyalov and Bourne, 1998, 2001) breaks the structures into series of domain fragments and derives optimal assignments of fragments called aligned fragment pairs (AFP) by means of rigid superposition. The solution is subsequently expanded by combinatorial extension to derive a global alignment of the structures. Its successor, CE-MC, uses the original CE approach to derive seed alignments that are subsequently expanded using Monte Carlo optimization (Guda et al., 2001, 2004).

Alignments based on SSEs can generally be expanded by using finer representations (e.g. atomic coordinates) in a second alignment step, subsequent to the SSE alignment.

This hierarchical alignment strategy virtually combines fold-based and geometric strategies and is used in a number of different approaches (Alexandrov and Fischer, 1996; Singh and Brutlag, 1997), the most prominent ones being VAST. Matras (MArkov TRAnsition of protein Structure evolution) uses Markov transition models to evaluate the alignments generated by a hierarchical alignment strategy that employs dynamic programming (Kawabata and Nishikawa, 2000).

MAMMOTH (MAMMOTH-mult) similarly decomposes the proteins into heptapeptides but uses dynamic programming to generate global alignments (Ortiz et al., 2002). More recent approaches are SABERTOOTH (Teichert et al., 2007), MISTRAL (Micheletti and Orland, 2009) and Fr-TM-Align (Pandit and Skolnick, 2008). SABERTOOTH condenses residue connectivity to structure profiles in vectorial form and calculates alignments using Dijkstra's shortest path algorithm (Dijkstra, 1959). MISTRAL calculates alignments by minimizing an energy function using simulated annealing and Fr-TM-Align expands a seed solution of aligned fragment pairs with respect to the TM-metric (Zhang and Skolnick, 2004) by using dynamic programming.

MUSTA (Leibowitz et al., 2001) and MultiProt (Shatsky et al., 2002b) are both geometric approaches to multiple structure comparison from the Nussinov-Wolfson group. MUSTA calculates multiple superpositions of complete proteins. It is more akin to template-based methods since it relies on detecting conserved patterns of amino acids (see Section 2.2.2). Multiprot calculates multiple structure alignments by precomputing sets of congruent fragments and subsequently derives local alignments of C_α atoms by means of a heuristic. The local alignments are then combined into a global solution. In this respect, Multiprot is special, as it essentially constitutes a semi-global approach.

Most of these approaches constrain themselves by relying on the sequential ordering of residues. The few non-sequential approaches include GANGSTA+, MISTRAL, MUSTA and Multiprot, although the latter strongly benefits from including sequential information.

Many global structure comparison methods, especially early approaches, rely on rigid structure comparison. Hence the most prominent evaluation criterion for structural alignments is the RMSD (Root Mean Squared Deviation) which is a measure of geometric deviation between superimposed alignments. The problem of calculating a structural alignment can thus be formulated as a minimization problem (minimizing the RMSD), a strategy realized by WHAT IF (Vriend and Sander, 1991). Other approaches aim to find the best superposition of two proteins by minimizing the surface between virtual protein backbones (Falicov and Cohen, 1996). However, these approaches typically do not account for molecular flexibility.

Until now, only few approaches exist that take molecular dynamics into account, such as Flexprot (Shatsky et al., 2002a, 2004), FATCAT (Ye and Godzik, 2003), TOPS++FATCAT (Veeramalai et al., 2008) and ALADYN (Potestio et al., 2010). These approaches typically account for flexibility by generating alignments consisting of rigidly aligned fragment pairs (AFP) interspersed with non-matching *hinge* regions. FATCAT and its successor TOPS++FATCAT use dynamic programming to derive complete alignments from sets of AFPs, the latter with a tremendous speed-up by pruning the search space of FATCAT using alignments of extended TOPS+ representations (Topology Of Protein Structures) (Veeramalai and Gilbert, 2008).

From a methodological point of view, many strategies are used to tackle the problem of structural alignments. Dynamic programming (Kawabata and Nishikawa, 2000; Ortiz et al., 2002; Pandit and Skolnick, 2008; Singh and Brutlag, 1997), double dynamic programing (Orengo and Taylor, 1996; Russell and Barton, 1992; Taylor et al., 1994) or iterative dynamic programming (Gerstein and Levitt, 1996, 1998) is often employed, but also graph-theoretic approaches are common (Gibrat et al., 1996; Krissinel and Henrick, 2004a; Madej et al., 1995; Redfern et al., 2007; Teichert et al., 2007), as well as geometric hashing (Dror et al., 2003; Leibowitz et al., 2001), genetic algorithms (Kolbeck et al., 2006), combinatorial optimization (Guda et al., 2004; Micheletti and Orland, 2009), three-dimensional clustering (Mizuguchi and Go, 1995) and other methods.

Global structure comparison can in principle recover evolutionary conservation that is not apparent at the sequence level. However, while often successful (Sánchez and Šali, 1997; Thornton, 2001; Zarembinski et al., 1998), similarity on the fold level does not always correspond to functional similarity, as cases are known where proteins with similar folds carry out different functions (Copley et al., 2004; Nagano et al., 2002; Orengo et al., 1999) as well as cases where the same function is carried out by proteins with different fold geometries (Polacco and Babbitt, 2006; Russell et al., 1998; Thornton et al., 1999; Wang and Samudrala, 2005). Moreover, surprising structural similarities can be missed when focusing on secondary structure elements alone (Jaroszewski et al., 2000). Another drawback of global protein comparison is the relatively huge computational cost associated, especially for higher levels of detail, which renders them less suited for large-scale analyses. As functional annotation cannot reliably be transferred based on global protein comparison (Rost, 2002; Todd et al., 2001; Wilson et al., 2000), local approaches can provide a more accurate view on functional similarity. Among these, one can in principle distinguish template-based from surface-based approaches.

2.2.2 Template-based structure comparison

The active sites of proteins are often more conserved than the overall fold, especially the three-dimensional arrangement of enzyme active site residues. The classic example is the catalytic triad of serine proteases, consisting of highly conserved apartate/glutamate, histidine, and serine residues. Hence, template-based methods focus on identifying conserved spatial arrangements of functionally important residues, such as catalytic dyads, triads or other catalytic centers. The detection of such structural motifs can be used to identify local functional similarities among proteins of different folds. Most of these approaches scan user-defined or automatically generated templates against a database of structures to detect frequent patterns.

One of the first approaches in this field is the ASSAM (Amino Acids Search for Substructures And Motifs) algorithm (Artymiuk et al., 1994; Spriggs et al., 2003), which employs graph theory to recover residue templates in the form of amino acid side chain patterns. Similarity of conserved patterns is established based on inter-atomic distances between the side chain atoms. Another graph-based approach which extracts templates purely based on the geometric information is DRESPAT (Wangikar et al., 2003). DRESPAT uses a clique-detection approach (Bron and Kerbosch, 1973) to derive pattern candidates and subsequently extracts conserved patterns from a set of proteins, using C_α and C_β atoms. SuMo (Jambon et al., 2003, 2005) similarly employs graph theory but uses triangles of stereochemical groups instead of the more widespread C_α representation and focuses on the similarity of physicochemical property.

Another commonly applied methodology for the detection of residue template is geometric hashing, which originates from the field of computer vision (Lamdan and Wolfson, 1988; Lamdan et al., 1988). The Nussinov-Wolfson group first employed the geometric hashing paradigm on protein data (Bachar et al., 1993; Fischer et al., 1994; Nussinov and Wolfson, 1991). In contrary to the ASSAM approach, their algorithm utilized a geometric model of proteins based on C_α atom representation. TESS also employs geometric hashing (TEmplate Search and Superposition) (Wallace et al., 1997), aiming at recovering similar substructures to a user-defined query template from a hash table compiled from the PDB (Berman et al., 2000). TESS was superseded by the more rapid JESS algorithm, which employs dynamic programming (Barker and Thornton, 2003). This approach is associated with the Catalytic Site Atlas (Porter et al., 2004), a database containing templates that are either experimentally determined or derived from homologues by PSI-BLAST, which can be queried directly using JESS.

These early approaches have the drawback of being relatively restrictive concerning the matching of similar substructures, at best permitting for some distance variance by allowing inter-atomic distances to deviate within a certain range. More tolerant approaches are the SPASM (SPatial Arrangement of Side chains and Main chain) algorithm (Kleywegt, 1999) and FFF (Fuzzy Functional Forms[1]) (Fetrow and Skolnick, 1998). The former approach uses C_α atom coordinates as well as side chain pseudoatoms (essentially the center of mass of a side chain) in a recursive, depth-first tree search, whereas the latter only considers C_α atoms and essentially relies on a literature mining approach to derive a set of structural constraints for the putative catalytic sites.

PINTS (Patterns In Non-homologous Tertiary Structures) (Stark and Russell, 2003) simply calculates the largest common three-dimensional arrangement of residues by using a recursive depth-first search algorithm based on string matching (Russell, 1998), but unlike other approaches derives a statistical measure of significance for the obtained patterns which allows to judge the results more carefully. MASH (Match Augmentation with Statistical Hypothesis testing) (Chen et al., 2007) again uses a graph-theoretic approach to retrieve seeds of matching patterns in a structure with respect to a search template. These seeds are subsequently refined using the so-called *geometric sieve* algorithm. Similar to PINTS, MASH optimizes template matching by minimizing the RMSD of matched atoms and calculates a statistical estimate of significance. LabelHASH is the successor of MASH and additionally incorporates geometric hashing as a preprocessing step (Moll and Kavraki, 2008).

Most template-based methods have the drawback of requiring a great deal of expert knowledge about the structures in question, as templates either have to be manually predefined, or are derived from literature mining and annotated information. The PDB-SiteScan approach for example uses annotated information stored in the PDB Site entries to generate residue templates (Ivanisenko et al., 2004). However, some approaches can recover conserved residue templates automatically, for example the above-mentioned DRESPAT algorithm.

[1]The term *fuzzy* is applied here in a general sense referring to *a priori* unclear descriptors. The method itself does not utilize fuzzy logic.

2.2.3 Surface-based structure comparison

Surface-based and template-based methods are closely related since both groups focus on finding common substructures in different proteins. The main difference is that template-based approaches are usually limited to small residue patterns, such as catalytic triads, whereas surface-based approaches consider much larger areas, such as complete ligand binding sites. Moreover, the above-mentioned template-based approaches usually consider the complete protein structure to identify shared patterns, surface-based methods are focused on residues exposed to the surface of the protein. Another benefit of surface-based approaches is that they typically make no assumptions on the location, number or orientation of functionally important residues, nor do they require prior knowledge of catalytic residue templates and are in general sequence- and fold-independent.

Surface-based methods usually seek to extract and compare protein binding sites, clefts on the surface of proteins where ligands and substrates can be bound, based on the notion that functionally important sites are usually located in such clefts. Clefts provide an increased surface to interact with the ligand and solvent may be excluded from the binding site, which increases the probability of forming hydrogen-bonds and hydrophobic interactions with small ligands. The active site of a protein is usually located in the largest cleft on the surface (Laskowski et al., 1996; Peters et al., 1996).

Approaches to surface-based protein analysis usually consist of three interrelated components: A method to detect and extract a binding site, a model to represent these sites and a method to compare them and score their similarity.

2.2.3.1 Extracting putative protein binding sites

For the extraction and modeling of binding sites, a few basic strategies can be distinguished. SURFNET (Laskowski, 1995) identifies binding sites by fitting spheres of different sizes between protein atoms to detect crevices within the protein. Three-dimensional surfaces are then calculated based on an electron density function. This is similar to the POCKET approach (Levitt and Banaszak, 1992) with the difference, that SURFNET places spheres between each pair of atoms, whereas POCKET places spheres on a regular lattice of grid points. PASS is another sphere filling approach (Brady and Stouten, 2000). SURFNET was later superseded by SURFNET-ConSurf (Glaser et al., 2006), which combines the original SURFNET approach with ConSurf-HSSP. ConSurf identifies functionally important residues by estimating phylogenetic conservation of residues and mapping it back to the surface regions of a protein (Armon et al., 2001). This was later coupled with the HSSP (Homology-Derived Secondary Structure of Proteins) database (Glaser et al., 2005). ConSurf-HSSP is used in a second filtering step after SURFNET to remove spheres that are too distant from evolutionary conserved residues.

Silberstein et al. (2003) computationally distribute organic solvent molecules (e.g. t-butanol, acetone, etc) on the protein surface to perform an *in silico* solvent mapping. The interaction energies between the molecules and the enzyme are optimized using a molecular mechanics function. Q-SiteFinder uses a van der Waals probe (methyl group) to calculate interaction energy between the probe and protein sites in order to find energetically favorable binding sites (Laurie and Jackson, 2005). PocketFinder[2] similarly employs a van der Waals probe but uses an aliphatic carbon as probe instead of a methyl group (An et al., 2005).

Grid-based approaches like LigSite (Hendlich et al., 1997) or PocketPicker (Weisel et al., 2007), embed proteins into a regular-spaced Cartesian grid. Grid points are assigned a degree of "burial" and binding pockets are identified as clusters of highly buried grid points. The surface of a binding pocket is usually approximated by the closest grid points outside the van der Waals radii of protein atoms. Especially LigSite has been used in a variety of applications including docking (Rarey et al., 1995), *de novo* drug design (Verdonk et al., 2001) and to detect putative protein binding sites for the CavBase database (Schmitt et al., 2002).

Another approach developed by Edelsbrunner *et al.* uses weighted Delaunay triangulation to calculate α-shapes and extracts pockets and voids[3] from them (Edelsbrunner et al., 1998; Liang et al., 1998a,b). Extracted pockets and voids must at least be large enough to contain a water molecule (Binkowski et al., 2003a) and are stored in the CASTp database (Binkowski et al., 2003b). A related approach that also utilizes alpha shapes detects functionally sites by searching for so-called "split pockets", binding sites that are split by a cognate ligand (Tseng and Li, 2009).

The recently published ConCavity algorithm of Capra et al. (2009) combines evolutionary conservation with structural methods to predict the binding sites of small ligands. ConCavity combines the output of purely structural algorithms (e.g. LigSite, SURFNET, PocketFinder) and combines them with a weighting scheme based on residue conservation (Capra and Singh, 2007).

[2]This algorithm is not to be confused with Pocket-Finder, which is basically an implementation of LigSite.

[3]Pockets correspond to cavities and thus have access to the solvent, voids are completely enclosed within the protein.

A major problem of these approaches, is their inaccuracy in determining the borders of the cavity, leading to different binding site representations even for the same protein. However, the extraction of protein binding sites is only the first step in surface-based structure analysis. The more important issue for this thesis is the comparison of binding sites. This is, of course, closely related to the way, how binding sites are represented.

2.2.3.2 Binding site representation and comparison

FEATURE (Bagley and Altman, 1995), another early approach, creates descriptions of the 3D microenvironment based on a variety of physicochemical properties, such as charge, hydrophobicity or certain chemical groups (e.g. amide groups, hydroxyl groups). In this case, sites have to be specified manually. While this is clearly a drawback, FEATURE represents one of the first approaches that aim to model important functional sites by physicochemical properties rather than the amino acid composition alone. The FEATURE representation is used by S-BLEST (Structure-Based Local Environment Search Tool) (Mooney et al., 2005), an algorithm that enables rapid database searches by comparing feature vectors of physicochemical attributes. Another more recent feature-based approach aims at the characterization of protein binding sites in terms of structural and physicochemical properties (i.e. size, shape, polarity, charge, electrostatics, flexibility, secondary structure, and hydrogen bonding capabilities, etc.). Overall, 408 features were used, thus representing the most exhaustive feature-based approach to protein binding site analysis so far. Principle Component Analysis (PCA) was used to extract the most discriminative features and the Euclidean distance of cavities in the PCA was used for clustering similar pockets (Andersson et al., 2010).

Mason et al. (2004) dissected binding pockets into smaller binding volumes that are represented by sets of contiguous cubic cells associated with physico-chemical descriptors via their QSCD (Quantized Surface Complementarity Diversity) method. These binding volumes are then mapped back to potential binding partners, which are modeled in the same reference frame.

The EF-site database contains information about the Connolly surface (solvent accessible surface area) (Connolly, 1983) of binding sites along with the electrostatical potential derived from Poisson-Boltzmann equations based on a precise continuum model (Nakamura and Nishida, 1987). Binding sites are represented as triangle meshes, with the vertices of each triangle labeled with electrostatic potential and curvature (Kinoshita and Nakamura, 2003, 2005). The comparison is done using a maximum clique detection approach (Bron and Kerbosch, 1973).

CavBase is a database for the automated detection, extraction and comparison of protein binding sites (Schmitt et al., 2001, 2002). Binding sites are extracted using the LIGSITE algorithm and represented using physico-chemical descriptors. As this thesis is based on the CavBase representation, a more detailed description is given in Chapter 4. A similar representation is used by SiteEngine (Shulman-Peleg et al., 2004), although both approaches employ different rules for the assignment of these properties. SiteEngine employs geometric hashing for the comparison of protein binding sites, whereas CavBase again utilizes graph theory.

The pvSOAR approach of Binkowski et al. (2003a) is built on the CASTp database and combines sequence information with spatial positioning of pocket-flanking residues. After detecting cavities using α-shapes, pocket-flanking residues are concatenated in sequence order and compared using dynamic programming. This comes at the price of being sensitive to the sequence order of residues, in other words, similar cavities with a different sequence order will not yield a high similarity score. Additionally, the shape similarity of the detected pockets is assessed by calculating the RMSD from a unit vector transformation based on a structural alignment (Umeyama, 1991). The pvSOAR approach was later expanded (pevoSOAR (Tseng et al., 2009)) by using individually derived evolutionary substitution matrices for searching similar patterns in CASTp instead of standard scoring matrices like BLOSUM50, which was used in the original approach. Of course, this requires the knowledge of different evolutionary related structures.

SiteBase uses an all atom representation of known ligand binding sites along with a pre-calculated similarity scores (Gold and Jackson, 2006). Binding site atoms are defined as atoms being within a certain cutoff range from co-crystallized ligands. Comparison is done using a geometric hashing approach (Brakoulias and Jackson, 2004) and similarity is measured by the number of matching atoms, but also sequence similarity. SOIPPA represents functional sites by Delaunay tessellations of C_α-atoms of the corresponding protein structures (Xie and Bourne, 2008). Each C_α-atom is labeled with a geometric potential, a measure based on the atoms distance to the protein surface and neighboring C_α-atoms (Xie and Bourne, 2007). Two protein sites are represented as graphs (based on the delaunay triangulation) and subsequently aligned by solving a maximum-weight common subgraph problem using a clique-detection approach (Östergård, 1999).

Najmanovich et al. (2007) used the SURFNET algorithm to extract protein binding sites and represent them in the form of graphs using C_α coordinates as nodes. Similarity is assessed both on the sequence level, using Hidden Markov Models (Eddy, 1998) and on the structural level, using a two-step graph matching procedure based on maximum clique-detection (Bron and Kerbosch, 1973). Structural similarity is measured by a local Tanimoto score which is basically the normalized size of the largest common subgraph.

The SURFACE database, a database containing annotated and compared surface regions from a large scale surface comparison study, similarly employs the SURFNET algorithm for the detection of largest clefts on the protein surface (Ferre et al., 2004). Cleft-flanking residues are subsequently selected and functional annotation is retrieved from the PROSITE database, which contains informations on protein domains, families and functional sites as well as associated patterns and profiles (Hulo et al., 2006). The comparison of these functional sites is done by finding and expanding seeds of similar residues with respect to evolutionary conservation based on Dayhoff substitution matrices (Dayhoff et al., 1978) and RMSD value.

Apparently, the matching of similar sites is mostly approached by graph theory and geometric hashing. Another interesting method called MultiBind comes from the field of computational geometry (Shatsky et al., 2006). This approach is build on the SiteEngine representation of binding sites (Shulman-Peleg et al., 2004) and computes multiple alignments of binding sites by solving a k-partite 3D matching problem. To this end they utilize geometric hashing combined with a subsequent greedy heuristic to calculate alignments with respect to a pivot structure.

SURF's UP is a surface-based approach that compares complete protein surfaces. In that respect, it is quite special and the only global surface-based approach to protein comparison. Comparing the complete surface of proteins is not easily done since they tend to be very diverse, especially in non-functional regions. SURF'S UP circumvents this problem by projecting amino acid properties (acidic, basic, hydrophobic and polar) onto a regular sphere (Sasin et al., 2007). A similar approach has also been applied to binding sites alone. SiteAlign represents binding sites by projecting certain physico-chemical and topological properties onto a sphere (Schalon et al., 2008).

Another group of approaches focus on the comparison of the volumetric shapes formed by protein binding pockets (Peters et al., 1996). Binkowski and Joachimiak (2008) compared the shapes formed by cavity residues using the Kolmogorov-Smirnov distance on probability distributions based on interatomic distances. The recent VASP approach uses a volumetric representation of binding sites and compares the shapes via *constructive solid geometry* (Chen and Honig, 2010).

2.3 Graph theory in bio- and chemoinformatics

As outlined above, graph theory plays a role in some of the above-mentioned approaches to protein structure comparison exist, though graph theory itself is of course not limited to protein structure comparison. In fact, graphs have been widely used in chemoinformatics for the comparison of chemical compounds as well. In a wider sense, graph theory offers a convenient framework for molecular structure comparison in general. In life sciences as well as chemistry, graph theory is especially appealing, since graphs represent a powerful data structure to model structured objects in a formal way. More precisely, graphs naturally support the modeling of geometric and topological information contained in data of molecular structures. A major advantage is the fact that by using graph representations, geometric information is transformed into a relational representation, thereby putting all structures in the same reference frame.

The modeling of molecular structure data by means of graphs has a long-standing tradition in chemoinformatics (Balaban et al., 1976), but also increasingly in bioinformatics for the modeling of protein structure data, regardless whether one is interested in a (global) fold detection (Borgwardt et al., 2005; Grindley et al., 1993), the detection of spatially conserved templates (Artymiuk et al., 1994; Jambon et al., 2003; Spriggs et al., 2003) or the comparison of binding pockets (Kinoshita and Nakamura, 2003, 2005; Schmitt et al., 2002). Moreover, even when graphs are not directly used to model protein geometry, they are often used in underlying comparison algorithms, for example in the SOIPPA approach (Xie and Bourne, 2008).

Hence, the comparison of graphs is an important task that relates to these fields of application. As will be seen in the following section, a large number of different approaches to the comparison of graphs exist and have been applied to protein structure comparison. Among these are methods based on exact graph matching (Alexandrov and Fischer, 1996; Artymiuk et al., 1994; Kinoshita and Nakamura, 2003, 2005; Madej et al., 1995; Schmitt et al., 2002; Spriggs et al., 2003; Xie and Bourne, 2008), inexact matching (Tian and Patel, 2008; Tian et al., 2007; Weskamp, 2007) and feature-based approaches (Borgwardt et al., 2005; Vishwanathan et al., 2010).

Another important field utilizing graph theory is systems biology. In systems biology, graphs are routinely used for the modeling of biological networks, such as protein interaction and signaling networks (Berg and Lässig, 2004; Xenarios et al., 2002), metabolomic networks (Kanehisa et al., 2004), phylogenetic networks (Huson and Bryant, 2006) or gene regulatory networks (Davidson et al., 2002). Again, the comparison of graphs plays an important role for the analysis of these networks, to compare complete networks, but more often to detect common subgraphs. This is relevant, for example, to identify shared biochemical pathway from metabolomic networks (Kanehisa et al., 2004).

In principle, the same mechanisms that are applied in graph-based structure comparison can be employed in this context. Since this can potentially involve querying large graphs, several approaches originating from the field of database research combine graph comparison with indexing strategies to reduce the search space prior to graph comparison. Again, some of them focus on exact matching (Shasha et al., 2002; Yan et al., 2004; Zhang et al., 2007) and more recently also on error-tolerant graph matching (Yan et al., 2005, 2006), though one has to mention that these earlier methods are not as flexible as the more recent SAGA (Tian et al., 2007) and TALE (Tian and Patel, 2008) approaches.

Finally, in chemistry, graph comparison is an important technique to compare chemical compounds, which are routinely modeled as graphs (Bunke, 2000; Bunke and Jiang, 2000). Here, exact methods usually involve the detection of common subgraphs, either the maximum common subgraph between two or more graphs (Bunke and Jiang, 2000; Raymond et al., 2002) or even a set of frequent common subgraphs, which leads to the related problem of frequent subgraph mining (Borgelt and Fiedler, 2008; Kuramochi and Karypis, 2007; Yan and Han, 2003). As with the fields above, also approximate methods (Bunke, 1999; Justice and Hero, 2006) and feature-based methods (Neuhaus and Bunke, 2006a; Ralaivola et al., 2005) are applied.

Graph comparison is a vital part in all these applications, especially protein structure comparison, which is the main application considered in this thesis. Therefore, an overview over methods for graph comparison is given in the following section.

2.4 Graph comparison

In this thesis, the problem of comparing protein structures in general and protein binding sites in particular is formalized as a graph comparison problem. At least, the comparison of graphs in general requires a distance or similarity measure. Graph matching can be viewed as a special case of graph comparison. In particular, graph matching tries to derive a correspondences between graphs in terms of an alignment of nodes of two or more graphs. This gives rise to a similarity measure as well. In this section, an overview of approaches to the comparison of graphs is given.

The problem of graph comparison has originated from the field of pattern recognition but has since then spread to other fields of research, including data mining, (kernel-based) machine learning, (structural) bioinformatics, and many others. Graph comparison and, more specifically, graph matching usually depends on the concept of similarity that is employed. Roughly, graph comparison approaches can be divided into a few main categories that give rise to different classes of algorithms:

- *Exact graph matching* is characterized by being edge-preserving, that is, nodes that are connected by an edge in one graph must also be connected in the other graph to be considered for matching. In this scenario, graphs are considered similar if they are isomorphic or fulfill the subgraph isomorphism property (cf. Chapter 3). This is closely related to the concept of the maximum common subgraph and related methods and especially relevant in the field of chemoinformatics (cf. Section 2.4.1).
- Depending on the application, graphs can be subjected to deformations thus rendering the first principle inadequate as it imposes too stringent constraints on the graph matching process. This is especially the case in bioinformatics when modeling molecular structure data. Thus, the matching process must be tolerant which is realized in *inexact graph matching*. Here, the edge-preserving property is not necessarily satisfied by matching nodes, although violations are usually penalized. Among other strategies, this category includes approaches based on the generic concept of an *edit distance*. According to this principle, two graphs are similar if a few modifications (edit operations) are sufficient to make the first one isomorphic to the second one.
- *Feature-based comparison* avoids the problem of deriving a correspondence between the graphs themselves at all by converting them into a feature representation. Similarity is then calculated based on these feature representations. While this allows for the application of a plethora of methods from the field of machine learning, the question arises how to derive a suitable feature representation, since this strategy inevitably leads to a loss of information (cf. Section 2.4.3).

Subsequently, a brief overview of algorithms to exact graph comparison is given (cf. Section 2.4.1), followed by a discussion of approximate graph matching strategies (cf. Section 2.4.2). For a more exhaustive review on graph matching, see (Conte et al., 2004). While the former principle has often been applied for molecular structure comparison, especially but not exclusively in chemoinformatics, the latter principle is intuitively well-suited for inherently noisy biological data. In addition, some recent feature-based approaches are discussed (cf. Section 2.4.3).

2.4.1 Exact graph matching

Exact graph matching typically employs the concepts of graph isomorphism and subgraph isomorphism (cf. Chapter 3) to determine the similarity of graphs, for which algorithms have been known for a long time (Read and Corneil, 1977). Closely related to subgraph isomorphism is the principle of detecting common subgraphs. On the one hand, this can involve detecting a number of different common subgraphs of two or more graphs, which is the basic idea behind frequent subgraph mining. This is realized by several different approaches designed to mine large graph databases of chemical compounds such as GSpan (Yan and Han, 2002), ClosedGraph (Yan and Han, 2003), MOSS (Borgelt et al., 2005), FFSM (Huan et al., 2003) or gFSG (Kuramochi and Karypis, 2007). On the other hand, the maximum common subgraph (MCS) between two graphs is usually of great interest, as it cannot only be used to obtain a measure of similarity of the graphs, but moreover identifies the region with the greatest correspondence. This strategy is of particular interest, since it is widely used for molecular structure comparison in the context of chemoinformatics, more precisely the comparison of chemical compounds (Bunke and Jiang, 2000). A related concept, the maximum common supergraph has also been applied in this field (Bunke et al., 2000). Many distance measures for graphs in the context of exact graph matching in fact utilize the size of the MCS (Bunke and Shearer, 1998; Wallis et al., 2001).

The MCS itself is not uniquely defined as the maximality of the subgraphs is either referring to the number of nodes or, alternatively, the number of edges is maximized. The former variant is known as maximum common induced subgraph (MCIS) and the latter as maximum common edge subgraph (MCES). Both have been used separately (Bunke and Shearer, 1998) as well as combined in a single distance measure (Fernández and Valiente, 2001). Moreover, some algorithms consider only connected subgraphs, while others allow for a disconnected MCS. An excellent review on the field of MCS algorithms employed in chemoinformatics is given by (Raymond and Willett, 2002).

Matching problems based on the concepts above are NP-complete (Garey and Johnson, 1979), except for the graph isomorphism problem, for which it is still unknown whether it belongs to NP (Köbler et al., 1994). Thus, exact graph matching techniques usually have exponential time complexity in the worst case. Polynomial algorithms are only known for special variants of graphs, e.g. trees (Aho et al., 1974) or planar graphs (Hopcroft and Wong, 1974).

Most algorithms that approach the graph and subgraph isomorphism problem are based on tree search strategies in conjunction with some sort of pruning procedure. For example, one of the first important algorithms to address the problem of graph and subgraph isomorphism uses depth-first search with backtracking, followed by a look-ahead procedure evaluating a matrix of possible future matches (Ullmann, 1976). Despite its age, this approach is still widely used, also in structural bioinformatics as the main algorithm behind ASSAM (Artymiuk et al., 1994; Spriggs et al., 2003). Another recent tree-based approach to both graph and subgraph isomorphism uses a heuristic considering the set of neighboring nodes to the ones already considered in a partial matching (Cordella et al., 2004). The nRF+ algorithm reformulates the subgraph isomorphism problem as a constraint satisfaction problem (CSP) and uses heuristics to obtain a solution (Larrosa and Valiente, 2002).

Finding the maximum common subgraph is the most prominent approach in the context of graph-based structure comparison. The MCS problem can be reformulated as a clique-detection problem, that is, finding the largest complete (fully connected) subgraph in a *product graph* (or alternatively *association graph*) (cf. Section 3) derived from the *factor graphs*[4]. While the Ullmann algorithm (Ullmann, 1976) can be used to solve the MCS problem as well by employing clique detection, other approaches are more efficient.

One of the first and the most widely used clique detection approaches is the Bron-Kerbosch algorithm (Bron and Kerbosch, 1973) which also employs a tree-search in combination with a backtracking strategy. More precisely, it constitutes a clique enumeration algorithm employing again a depth-first search. Despite its age, the Bron-Kerbosch approach is still the core algorithm of many structural comparison approaches, including the above-mentioned SOIPPA (Xie and Bourne, 2008), EF-site (Kinoshita and Nakamura, 2005) and CavBase (Schmitt et al., 2002). A similar approach uses a heuristic based on graph coloring to explore the product graph (Balas and Yu, 1986). Later approaches tried to speed up the clique enumeration process, e.g., by using information from the factor graphs (Bessonov, 1985) while others focus on connected MCS only[5] (Koch, 2001; Tonnelier et al., 1990).

Other algorithms solve the MCS problem directly without converting it to a clique-detection problem, employing again backtracking (McGregor, 1982; Schmidt and Druffel, 1976). However, due to the advance in computer technology, these approaches are rendered less efficient than clique-detection approaches today. The algorithm of Akutsu solves the MCS problem via dynamic programming (Akutsu, 1993). Another group of

[4]The input graphs being compared.

[5]The previous approaches allow for disconnected cliques.

algorithms utilize preprocessing of the graphs to allow for a faster comparison. For example, some algorithms pre-compute a canonical labeling to allow for the application of hashing methods (McKay, 1981; Messmer and Bunke, 2002). However, the former algorithm, Nauty, only deals with graph isomorphism, limiting its use for structure comparison. Messmer and Bunke approached the isomorphism and subgraph isomorphism problem using a decomposition approach (Messmer and Bunke, 2002).

Also precomputed decision trees are employed, either to compute the matching directly for the isomorphism and subgraph isomorphism problem (Bunke and Messmer, 1997; Messmer and Bunke, 1999), the MCS problem (Shearer et al., 2001) or to use it as a quick pre-filtering step in combination with a complete matching procedure (Irniger and Bunke, 2001; Lazarescu et al., 2000).

A problem of exact matching algorithms is the high runtime complexity. One possible strategy to mitigate this problem is to employ heuristics in order to calculate solutions within tolerable time complexity. This obviously comes at the price of possibly suboptimal solutions that might not even be close to the global optimum. Especially the MCS problem has been approached in this manner by casting the MCS problem to a continuous optimization and applying different optimization techniques, such as greedy optimization (Funabiki and Kitamichi, 1999), evolutionary optimization (Brown et al., 1994; Wagener and Gasteiger, 1994; Wang and Zhou, 1997), simulated annealing (Barakat and Dean, 1991) and neural networks (Schädler and Wysotzki, 1997). Heuristics based on graph walking procedures have been employed as well (Bayada et al., 1992; Hagadone, 1992). Alternatively, the complexity problem can be mitigated by parallel computing (Pardalos et al., 1998; Shinano et al., 2002).

Another major drawback of exact graph matching is the relative stringent definition of similarity, rendering them less suited for the application on noisy data, including real geometric protein structure data. The use of approximate or inexact graph matching techniques might help to alleviate this problem.

2.4.2 Inexact graph matching

As mentioned above, when modeling noisy real data, one has to account for a certain variability, either by employing a suitable model or by using an inexact comparison approach. In the latter case, one cannot always expect exact matches, especially for experimentally determined molecular structure data. Moreover, even when capturing this noise on the model level, there might still be an underlying variability in the patterns one is interested in. Hence, *approximate* or *error-tolerant* graph matching techniques are required.

For approximate graph matching approaches, the edge preservation property does not necessarily hold for all matched nodes. For this, typically a penalty is used, e.g. in the context of a scoring scheme. Approaches that derive a scoring scheme on an explicit error model are usually referred to as *error-tolerant* or *error-correcting* graph matching. An alternative way to obtain a cost function is to allow certain types of edit operations (typically insertions, deletions, substitutions or label/weight changes of nodes/edges) to which a certain cost is assigned. The *graph edit distance* is then given by the minimal sequence of edit operations needed to transform one graph into the other. The graph edit distance can be regarded as a distance function on graphs and a first definition was published by (Sanfeliu and Fu, 1983), although it could be shown that their distance function is not a metric (Shapiro and Haralick, 2009).

Using the graph edit distance to derive a graph matching is is a more general approach than the above-mentioned exact approaches. It could be shown that (sub)graph isomorphism and especially the MCS problem are special instances of the graph edit computations (Bunke, 1999). Moreover, the GED can be used to evaluate matchings derived by different means. Wang and Wang (2000) used the GED to evaluate approximate matchings of chemical compounds derived by the well-known geometric hashing paradigm (Lamdan and Wolfson, 1988).

So basically, inexact graph matching corresponds to an optimization problem with respect to a certain cost function. While a variety of different algorithms exist to approach this sort of problem, many fall into one of two major categories:

- Tree-based approaches, as utilized for the exact matching problem, are also widely used for the inexact case. Typically, cost functions are used to guide the tree search.
- A large group of algorithms translates the inexact matching problem, which is basically a discrete optimization problem, into a continuous problem.

Many of the first inexact graph matching algorithms utilize tree search guided by the costs incurred by assigned mappings. The first approach only considered node/edge substitutions (Tsai and Fu, 1979), while a later expansion allowed also for the matching of graphs with differing topology by include insertions and deletions (Tsai and Fu, 1983). An error-correcting algorithm for hypergraphs was proposed by Shapiro and Haralick (1981), other approaches utilize the A* algorithm to compute a graph distance (Berretti et al., 2002; Dumay et al., 1992; Gregory and Kittler, 2002).

Continuous methods transfer the discrete optimization problem into a continuous optimization problem, thus making it amenable to a large number of methods from continuous optimization. A first group of continuous optimization approaches employ relaxation labeling in a probabilistic framework (Kittler and Hancock, 1989). A later approach based on this framework allowed the use of node/edge attributes in the labeling update process (Christmas et al., 1995). To obtain a distance measure, the probabilistic framework of the Hancock group was expanded by the definition of a Bayesian consistency measure (Huet and Hancock, 1999; Wilson and Hancock, 1997). Myers *et al.* later introduced the definition of a Bayesian graph edit distance based on this framework (Myers et al., 2000).

Another group of algorithms focus on the weighted graph matching problem. Given two graphs with weights on edges, the weighted graph matching problem searches for an optimal permutation of nodes of one graph so that the difference between the edge weights is minimized. This problem has been approached using the simplex algorithm (Almohamad and Duffuaa, 1993), Lagrangian relaxation networks (Rangarajan and Mjolsness, 2002) and graduated assignment (Gold and Rangarajan, 1996). A simplified version of the problem has also been approached using fuzzy logic (Medasani et al., 2002). However, the problem of these methods is the fact that they are not applicable to node labeled graphs.

A general disadvantage of these continuous methods is that they do not guarantee to find an optimal solution. Moreover, the solution for the continuous problem must be converted back to a discrete case, which introduces another source of approximation. Hence, performance can vary, depending on the graphs and the application at hand and for specific problems they might produce relatively poor results.

Beyond these categories, still more algorithms exist. Messmer and Bunke (1998) extended their decomposition approach to (sub)graph isomorphism mentioned above to the inexact case as well, combining the GED with enumeration techniques and indexing methods. Other algorithms employ genetic algorithms (Cross et al., 1996; Wang et al., 1997), simulated annealing (Jagota et al., 2000; Xu and Oja, 1990), tabu search (Williams et al., 1999) or neural networks (Suganthan, 2002; Suganthan and Yan, 1998). A recent approach combines exact matching, namely the Bron-Kerbosch algorithm to clique detection, with a greedy heuristic to expand the original solution (Weskamp, 2007). Another more recent approach uses binary linear programming to calculate graph matchings based on graph edit distances (Justice and Hero, 2006).

While inexact graph comparison might offer some benefits for the analysis of noisy data and the search for variable patterns, one has to note that the performance of such methods strongly depend on the parameterization of the cost function (Bunke, 1999). The learning of a cost function is a problem on its own, that has been approached using for example the expectation maximization (EM) algorithm (Neuhaus and Bunke, 2004) or self-organizing maps (Günter and Bunke, 2002; Neuhaus and Bunke, 2005). As another downside, these algorithms suffer from the danger to get stuck in local optima, as they do not guarantee to find a global minimum of the matching cost. Also, one has to keep in mind that many of the above-mentioned approaches are only suitable for certain types of graphs and data. Most of the above-mentioned approaches only deal with unlabeled graphs or are even restricted to special kinds of graphs (e.g. planar graphs, trees, directed acyclic graphs, etc.). Another problem is again the time complexity, despite the application of parallel computing for inexact graph matching to speed up computations (Allen et al., 2002).

2.4.3 Feature based approaches to graph comparison

In principle, it is always possible to derive a feature representation of a graph and reduce the problem of comparing graphs to a feature comparison problem (Papadopoulos and Manolopoulos, 1999). While this makes them amenable to a huge number of machine learning methods, one has to keep in mind that this inevitably leads to a loss of information. For example, in its most simple case, one could represent graphs by a few key attributes, such as the cardinality of the node/edge set or the number of connected components. Obviously, in doing so, one would loose any information about the topology of the graph, the comparison, however, would be much simplified. Instead of using such general descriptors, a feature representation can be derived by using local features of a graph, hence approaches of that kind are termed local approaches in a methodological sense.

Local approaches to graph comparison generally look for the compliance of properties that refer to substructures or local components of a graph, such as subgraphs, paths or walks. In contrast to subgraph isomorphism approaches, local methods typically exploit sets of characteristic common substructures for a given group of graphs to derive a similarity measure, rather than a single maximum common subgraph. Reducing the graph comparison to the comparison of local features thus leads to a local similarity measure, in contrast to similarity measures based on graph isomorphism, subgraph isomorphism or MCS which are classical global similarity measures.

Main contributions to graph comparison based on this principle have recently been made on the field of kernel-based machine learning (Shawe-Taylor and Cristianini, 2004).

A kernel is a function defined on a set of complex objects satisfying certain mathematical properties which makes them appealing both from a mathematical and an algorithmic point of view (cf. Chapter 3). To put it simply, besides other benefits, kernel functions can also be regarded as similarity functions on complex objects, such as graphs.

Several different kernel functions on graphs have already been introduced in the context of the R-convolution framework which is based on the idea that the comparison of two complex objects can be reduced to an all-to-all comparison of their constituents (cf. Chapter 4). Basically, these kernel functions constitute local similarity measures as they focus on (local) subpatterns within a graph structure, such as walks or paths. These can be used for graph comparison, e.g., in classification scenarios. Kernels have initially been defined on subparts of a graph, e.g., node kernels such as the diffusion kernels (Kondor and Lafferty, 2002; Lafferty and Lebanon, 2003; Smola and Kondor, 2003) or kernels on paths within a graph (Takimoto and Warmuth, 2003). Among the first graph kernels defined *between* graphs were kernels based on random walks (Gärtner, 2003; Kashima et al., 2003). The basic idea is that, the more similar two graphs are, the more often a randomly generated walk from one graph will also be present in the other.

Originally, the kernels of Gärtner were only defined on node-labeled graphs. While the geometric kernel only considers node labels of the first and the last node of a walk (Gärtner, 2002), the random walk kernel compares node labels of the complete walk (Gärtner, 2003). This was later expanded to edge labeled graphs as well (Borgwardt et al., 2005).

Kashima *et al.* proposed to apply the concept of marginalized kernels (Tsuda et al., 2002) to graph kernels as well (Kashima et al., 2003, 2004), again focusing on random walks but representing them in terms of a sequence of node and edge labels. This was later extended by a relabeling of node labels to include further information about the local graph topology (Mahé et al., 2004). A problem of random walk kernels is the fact that the number of possible random walks can become quite large. Thus, as an alternative, shortest path kernels were proposed that utilize only the shortest path between nodes (Borgwardt and Kriegel, 2005).

Another possibility is to combine graph matching concepts with the definition of kernel functions. Neuhaus and Bunke used the graph edit distance defined above to derive a random walk kernel (Neuhaus and Bunke, 2006b) and a convolution kernel (Neuhaus and Bunke, 2006a) based on the GED. The former is basically an extension of the original random walk kernel (Gärtner, 2003), enhanced by restricting the random walks to those that satisfy certain global matching constraints as defined by an edit path. The latter is a convolution kernel comparing graph decompositions that can be transformed into each other by a valid graph edit path.

Recently, a function based on the pointwise inner product (Schur-Hadamard) was employed in a graph-based classification scenario, although this function does not fulfill all kernel properties (Jain et al., 2005). In the field of chemoinformatics, the optimal assignment kernel has recently been introduced that represents an alternative to the R-convolution framework. Given a kernel function defined on subcomponents of a graph, the optimal assignment kernel searches for an assignment of subcomponents by maximizing the sum over all mutually assigned components (Fröhlich et al., 2005). As with the Schur-Hadamard kernel, the optimal assignment kernel is mathematically not a real kernel (Vert, 2008). However, a variant of this method was recently proposed that provably fulfilled the kernel properties (Vishwanathan et al., 2010).

More kernels introduced for the comparison of chemical compounds were introduced in (Ralaivola et al., 2005). Those resemble certain concepts of molecular fingerprints of chemical compounds, namely the Tanimoto kernel, the min-max kernel and the hybrid kernel. An advantage of kernel functions is the fact that they can be computed relatively efficiently in general. Moreover, since they fulfill certain mathematical properties, they can be used to make graphs amenable to certain machine learning algorithms, such as support vector machines.

Aside from kernel-based approaches, alternatives exit that build upon different decomposition techniques. One major line of work focuses on spectral methods. Spectral methods are based on the notion that the eigenvalues and eigenvectors of a graph are invariant to the node ordering. Therefore, isomorphic graphs will have identical eigenvalues and eigenvectors, although the opposite is not true. Originally suggested by (Umeyama, 1988), some approaches try to exploit this principle for graph matching, using for example gradient descent (Xu and King, 2001) or clustering (Kosinov and Caelli, 2009). Another approach used spectral serialization to compute the graph edit distance (Robles-Kelly and Hancock, 2005). Recently, Kondor and Borgwardt introduced a set of invariant matrices derived from graphs by Fourier transformation called the skew spectrum (Kondor and Borgwardt, 2008). It could be shown that it could compete with state-of-the-art graph kernels. As the computation of eigenvalues is a well-studied problem, this idea is relatively popular, although these approaches are only of limited use, as they cannot use any node or edge label information.

Feature-based approaches offer in principle the possibility to reduce the complexity of the graph comparison problem, thus promising a much higher runtime efficiency. The downside of course is the loss of information inevitably incurred. As long as one can catch the important properties of a graph, this is not necessarily a problem. However, this is not easily done, especially since one main reason to use the more expressive graphs instead of a feature representation is the lack of a concrete set of discriminating features.

3

Preliminaries

The algorithms presented in Chapter 4 essentially constitute different approaches to graph comparison and as such focus on graph-based representations of protein binding sites. Thus, it is necessary to introduce some important key graph-theoretical concepts that apply to all of these methods. The CavBase database will provide the data basis for the upcoming experiments as well as the basic binding site representation that will be the background for the derived graph models. Again, this serves as the foundation for all presented methods.

In the experimental part, it will be necessary to obtain a measure of confidence for obtained similarity scores to interpret the results. This will also be addressed below. In the following, the stage will be set for the understanding of the subsequent chapters by briefly introducing these general aspects prior to the actual introduction of the developed methods.

3.1 Graph-theoretic foundations

In the following section, some graph-theoretic prerequisites and notations used throughout this thesis will be formally introduced.

3.1.1 An introduction to basic graph concepts

In the context of this work, protein structure comparison, or more precisely, protein binding site comparison, is reformulated as a graph comparison problem. More specifically, graphs are used to model structure data to make them amenable to graph-based algorithms. Mathematically, a graph can be defined as a tuple $G = (V, E)$, with V denoting a set of vertices, or nodes, and $E \subseteq V \times V$ denoting a set of edges connecting these nodes.

Definition 1 *(Graph)*
A graph is a tuple $G = (V, E)$, with V representing a finite set of vertices, or nodes, and $E \subseteq V \times V$ denoting a set of edges. Two nodes $v \in V$ and $u \in V$ are connected by an edge if $(v, u) \in E$. The cardinality of the graph is given by $|V|$.

A graph in general can include loops, that is, edges connecting one node with itself or even multiple edges between two nodes. In the latter case, the graph is also called a multigraph. A generalization of this concept, a *hypergraph*, can even contain edges between more than two vertices. However, throughout this thesis, the term graph is used with the implicit understanding that only *simple graphs* are considered.

Definition 2 *(Simple graph)*
A graph $G = (V, E)$ is called a simple graph, if the following conditions hold:

- $\forall v_i \in V : (v_i, v_i) \notin E$ *(no loops are present)*

- $\forall v_i, v_j \in V : |\{(v_i, v_j) | (v_i, v_j) \in E\}| \leq 1$ *(no multiple edges occur)*

In the context of this work, the terms *graph* and *simple graph* are used synonymously.

For the modeling of molecular structure data, this standard definition of a graph is insufficient as the graph model should also encode certain properties of nodes and edges defined by the underlying components of the structure. For example, suppose a node represents an amino acid within a protein structure. In that case, the amino acid type may be encoded in terms of a node label. Moreover, since the graphs will be used to represent geometric data, it will be necessary to include information about the distances between nodes, respectively structure components. Thus, undirected node-labeled and edge-weighted graphs will be used.

Definition 3 *(Node-labeled edge-weighted graph)*
Let Σ be a set of node labels. A node-labeled and edge-weighted graph G is a 4-tuple $G = (V, E, l, w)$, with V denoting a set of nodes and $E \subseteq V \times V$ denoting a set of edges. Additionally, a labeling function $l : V \rightarrow \Sigma$ and a weighting function $w : E \rightarrow \mathbb{R}^+$ is defined.

l is a labeling function that assigns to each node $v \in V$ a label in Σ and w is a weighting function, that assigns a non-negative weight to an edge $(v,u) \in E$[1].

Based on this definition, molecular structures can be modeled and compared using graph comparison techniques, including exact graph matching, inexact graph matching and feature-based approaches (cf. Chapter 2). It should be noted, that such a graph model is not necessarily restricted to the modeling of binding sites. In principle, any molecular structure can be modeled, accordingly.

As mentioned previously, a characteristic trait of exact graph matching techniques is the requirement to derive an edge-preserving matching in the sense that two nodes connected by an edge in the first graph may only be mapped to nodes in the second graph, if they are connected by an edge as well. In its most stringent form, this leads to the *graph isomorphism* concept. For undirected node-labeled and edge-weighted graphs, this is defined in the following way:

Definition 4 *(Graph isomorphism)*
Let $G_1 = (V_1, E_1, l_1, w_1)$ and $G_2 = (V_2, E_2, l_2, w_2)$ be undirected node-labeled and edge-weighted graphs. A graph isomorphism *is a bijection $f : V_1 \rightarrow V_2$ that satisfies the following criteria:*

1. $(u,v) \in E_1 \iff (f(u), f(v)) \in E_2$,

2. $l_1(v) = l_2(f(v)) \; \forall v \in V_1$,

3. $w_1(u,v) = w_2(f(v), f(u)) \; \forall (u,v) \in E_1$.

G_1 and G_2 are called isomorphic, denoted by $G_1 \cong G_2$, if such a bijection exists.

For graph isomorphism, the edge-preserving condition must hold in both directions and a mapping of two graphs must be bijective, establishing a one-to-one correspondence between each node of the first and each node of the second graph.

As the graph-theoretical approaches introduced in this work will be applied to node-labeled and edge-weighted graphs derived from protein structure data, a more relaxed

[1] Since the graphs in this thesis are undirected, it would be more correct to use a subset notation instead of a tuple notation. For convenience, the widely used tuple notation will be used, with the implicit understanding that $(v,u) \in E \iff (u,v) \in E$ and $w(u,v) = w(v,u)$

isomorphism concept will also be needed to account for a certain degree of variation among the edge weights. As will become clear later, the edge weights of the graph models represent the Euclidean distance between certain points within the modeled molecular structures (e.g., atoms or pseudocenters, see Section 3.2). Edge weights are derived from atom coordinates of experimentally determined structures and hence subjected to inaccuracies due to measurement errors, molecular flexibility and low resolution of the crystal structures. Thus, a certain tolerance with respect to edge weight deviations is required. To this end, the isomorphism concept introduced above will be altered to allow edge weights to deviate up to a certain threshold ε.

Definition 5 *(ε-Isomorphism)*
Let $G_1 = (V_1, E_1, l_1, w_1)$ and $G_2 = (V_2, E_2, l_2, w_2)$ be undirected node-labeled and edge-weighted graphs. An ε-isomorphism is a bijection $f : V_1 \rightarrow V_2$ that satisfies the following criteria:

1. $(u,v) \in E_1 \iff (f(u), f(v)) \in E_2$,

2. $l_1(v) = l_2(f(v)) \; \forall v \in V_1$,

3. $|w_1(u,v) - w_2(f(v), f(u))| \leq \varepsilon \; \forall (u,v) \in E_1$.

G_1 and G_2 are called ε-isomorphic, denoted by $G_1 \cong_\varepsilon G_2$, if such a bijection exists.

A weaker concept than graph isomorphism is *subgraph isomorphism*, for which the isomorphism must only hold between one graph and a node-induced subgraph of another.

Definition 6 *(Subgraph)*
Let $G = (V, E, l, w)$ be a graph, then $G_{sub} = (V_{sub}, E_{sub}, l, w)$ is a subgraph *of G if $V_{sub} \subseteq V$ and $E_{sub} \subseteq E \cap (V_{sub} \times V_{sub})$. If, in addition $E_{sub} = E \cap (V_{sub} \times V_{sub})$, then G_{sub} is called an* induced subgraph.

Definition 7 *(Subgraph isomorphism)*
A subgraph isomorphism between two graphs G_1 and G_2 exists, if G_{sub} is a subgraph of G_1 and $G_{sub} \cong G_2$ (or vice versa).

Closely related to the concept of subgraph isomorphism is the common subgraph, especially the *maximum common subgraph* (MCS). As outlined in Chapter 2, this is one of the most widely used concepts in exact graph matching when comparing molecular structure data, including the comparison of protein binding sites as used in the CavBase database. However, the MCS is not uniquely defined, as maximality can refer to the number of nodes as well as the number of edges. The former variant, which is more precisely termed "maximum common *induced* subgraph" (MCIS) is more frequently used and referred to as MCS, the latter is known as maximum common *edge* subgraph (MCES).

Definition 8 *(Maximum common induced subgraph)*
Let G_1 and G_2 be two graphs. G_{CS} is a common induced subgraph, if induced subgraphs G^1_{sub} of G_1 and G^2_{sub} of G_2 exist with $G^1_{sub} \cong G^2_{sub} \cong G_{CS}$.

Again, to account for some edge weight tolerance, ε-isomorphism is used instead of the isomorphism criterion to define the MCS for the purpose of comparing molecular structures in this work:

Definition 9 *(Maximum common induced ε-subgraph)*
Let G_1 and G_2 be two graphs. G_{CS} is a common induced ε-subgraph, if induced subgraphs G^1_{sub} of G_1 and G^2_{sub} of G_2 exist with $G^1_{sub} \cong_\varepsilon G_{CS}$ and $G^2_{sub} \cong_\varepsilon G_{CS}$.

In the remainder of this thesis, the term maximum common subgraph (MCS) is used synonymously with the maximum common induced ε-subgraph for convenience, as this is the only variant that will be used. The cardinality of a graph will always refer to the number of nodes, e.g., $|G| = |V|$.

As mentioned in Chapter 2, the MCS problem can be solved by reformulating it as a clique-detection problem. A clique is defined as follows:

Definition 10 *(Clique)*
Given a graph $G = (V, E)$, a clique $G_q = (V_q, E_q)$ is a complete induced subgraph of G, so that $(u, v) \in E_q$ for all $u, v \in V_q$. A graph is called complete, *if each node is connected to every other node, i.e. the degree of each node is equal to the cardinality of the node set V minus 1: $\forall v_i \in V : deg(v_i) = |V| - 1$.*

The degree of a node v is the number of nodes $v_j \in V$ that are adjacent to v, in other words, that are connected to v via an edge:

$$deg(v_i) = |\{(v_i, v_j) | v_i, v_j \in V, (v_i, v_j) \in E\}| \tag{3.1}$$

If each node of a graph has the same degree it is also called a *regular graph*. The maximal clique is simply the clique with the highest cardinality:

Definition 11 *(Maximal clique) Let* $\mathbb{G}_{\shortparallel}$ *denote the set of all cliques of* G*, then* $G_{max} \in \mathbb{G}_{\shortparallel}$ *with* $|G_{max}| \geq |G_q|$ *for all* $G_q \in \mathbb{G}_{\shortparallel}$ *is a maximal clique.*

Obviously, every complete graph is identical to its own maximal clique.

To calculate the MCS, the product graph (or association graph) of two graphs is constructed and used as input for clique detection algorithms. Keeping the relaxation of the stringent isomorphism criterion in mind, the product graph in this work will be defined as follows:

Definition 12 *(Product graph, association graph)*
Given two graphs $G_1 = (V_1, E_1, l_1, w_1)$ *and* $G_2 = (V_2, E_2, l_2, w_2)$*, the product graph* $G_\times = (V_\times, E_\times)$ *is defined by its set of nodes* $V_\times \subseteq V_1 \times V2$ *and its set of edges* $E_\times \subseteq V_\times \times V_\times$ *satisfying*

1. $V_\times = \{(v_1, v_2) | v_1 \in V_1, v_2 \in V_2, l_1(v_1) = l_2(v_2)\}$

2. $E_\times = \{((v_1, v_2), (v'_1, v'_2)) \mid (v_1, v_2) \in V_\times, (v'_1, v'_2) \in V_\times, |w_1(v_1, v'_1) - w_2(v_2, v'_2)| \leq \varepsilon \vee ((v_1, v'_1) \notin E_1 \wedge (v_2, v'_2) \notin E_2)\}$

The nodes of a product graph consist of pairs of nodes from the two original input graphs, termed *product nodes*, connecting edges are respectively called *product edges*. A product edge exists, if either the difference between both factor edges are lower than ε, or if in both graphs the edges do not exist. The maximal clique of the product graph then corresponds to the MCS of the two original graphs (Levi, 1973). This notion is the basis for the Bron-Kerbosch algorithm.

Clique detection is a core component of the previously proposed algorithm by Weskamp (2007), as well as the internal comparison approach of CavBase (Schmitt et al., 2002). In the next chapter, it will be seen that this concept plays a major role in some of the introduced approaches in this thesis as well.

Furthermore, two other concepts from graph theory will be needed, the concept of walks and paths. In graph theory, a walk is defined as follows:

Definition 13 *(Walk)*
A walk w in a graph $G = \{V,E\}$ is a sequence of nodes $w = v_1, ..., v_{n+1}$ such that $(v_i, v_{i+1}) \in E$ for all $1 \leq i \leq n+1$. The length of a walk is given by the number of edges n.

Obviously, a walk represents an arbitrary sequence of egde-connected nodes, which means that nodes can occur multiple times in a walk. This is in contrast to a (simple) path, which is defined as follows:

Definition 14 *(Path, simple path)*
A path p in a graph $G = \{V,E\}$ is a sequence of nodes $p = v_1, ..., v_{n+1}$ such that $(v_i, v_{i+1}) \in E$ for all $1 \leq i \leq n+1$ and $v_i \neq v_j$ for all $v_i, v_j \in p$.

In other words, a path is a walk were no two nodes are identical. In the literature, this is also referred to as simple path.

3.1.2 The concept of graph alignments

Several approaches presented in this thesis will be designed to derive an overall correspondence between graphs, respectively the modeled protein binding pockets. This is realized in terms of graph alignments, which can be viewed as the graph-based counterpart to the concept of sequence alignment.

In case of graph alignments, a one-to-one correspondence of constituents naturally translates to a mutual assignment of nodes of the compared graphs. This idea has already been successfully applied in the context of structural bioinformatics, e.g., to derive common residue patterns in the ASSAM approach (Artymiuk et al., 1994) or the comparison of protein binding sites derived from CavBase (Weskamp, 2007). In the context of this work, the definition of a graph alignment is adopted from the latter work. There, the concept was defined for the more general multiple case, which will be also introduced here, though for the remainder of this work, the focus will be on the pairwise case.

Similar to sequences, protein structures vary in size and are subjected to mutations that result in insertions, deletions or alterations of amino acids. Molecular structures in general vary in size, which implies that in order to compare such structures by means of alignments, one has to account for the possibility of gaps, simply as nodes in one graph might not have a correspondence in another. Thus, the possibility of matching a node to a gap is introduced, with $\perp$ denoting gaps in the alignment.

Formally, a multiple graph alignment can now be defined as follows:

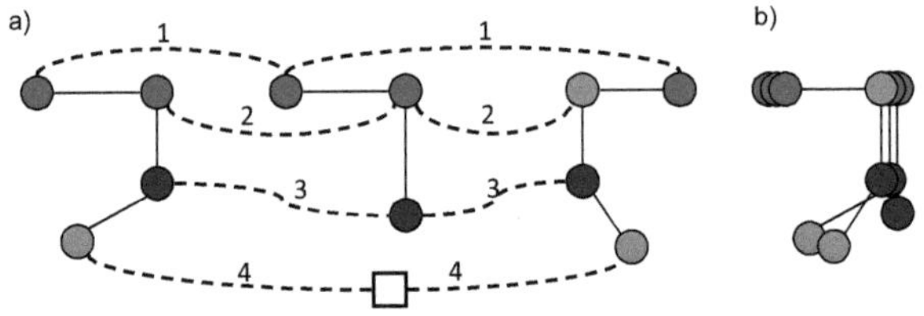

Figure 3.1: a) A valid multiple graph alignment for three distinct graphs. The node label is indicated by the coloring of the nodes, dashed lines indicate the assignment of nodes and the square denotes a gap. b) An overlay of the aligned graphs.

Definition 15 *(Multiple Graph Alignment)*
Let $\mathbb{G} = \{G_1(V_1, E_1), \ldots, G_m(V_m, E_m)\}$ *be a set of edge-weighted node-labeled graphs and let* $\perp$ *denote a gap in the alignment. Then*
$\mathscr{A} \subseteq (V_1 \cup \{\perp\}) \times \cdots \times (V_m, \{\perp\})$ *is an alignment of the graphs in* $\mathbb{G}$ *if and only if the following two criteria hold:*

1. *For all* $i = 1, \cdots, m$ *and for each* $v \in V_i$ *exists exactly one* $a = (a_1, \ldots, a_m) \in \mathscr{A}$ *such that* $v = a_i$.

2. *For each* $a = (a_1, \ldots a_m) \in \mathscr{A} : |\{a_i | a_i \neq \perp\}| \geq 1$.

Each $a \in \mathscr{A}$ represents a vector within the alignment that contains all mutually assigned nodes of the graphs $G_1, \ldots, G_m$. Note that by deriving a mutual assignment of nodes, edges are also implicitly assigned to each other. Hence the above definition completely defines a valid graph alignment.

The first condition enforces that each node v of any graph $G_i \in \mathbb{G}$ occurs exactly once in the alignment, ensuring that no node is mapped to more than one node in another graph. The second condition excludes vectors that consist exclusively of gap positions. An example of a valid graph alignment is depicted in Fig. 3.1.

While this general definition addresses the alignment of multiple graphs, the main focus in this work will be on the pairwise case. However, for those approaches able to generate a pairwise graph alignment, this is not a limitation per se, since pairwise graph alignments can easily be aggregated to a multiple graph alignment. In fact, this has also been done in the original publication, where initially only pairwise alignments were calculated and subsequently combined to a multiple one (Weskamp, 2007).

Viable aggregation strategies can again be adopted from the realm of sequence alignments, e.g., in the form of star alignment (Gusfield, 1993). The star alignment strategy known from multiple sequence comparison (Böckenhauer and Bongartz, 2007) constructs a multiple sequence alignment from pairwise alignments by choosing one sequence as a pivot element and aligning all other sequences relative to the pivot element.

The same strategy can also be applied to graph alignments. By selecting one graph as a pivot graph, one can align the nodes of all other graphs based on pairwise graph alignments with the pivot graph. More precisely, assume that without loss of generality a graph G_1 is selected as a pivot graph from a set of graphs $\mathbb{G} = \{G_1, \ldots, G_m\}$ to be aligned. If a node $v_1 \in V_1$ is mapped independently to nodes $v_2 \in V_2, v_3 \in V_3, \ldots, v_m \in V_m$ in separate pairwise alignments, these nodes are also be mapped onto each other in a multiple alignment. Hence, a tuple containing these nodes should be part of the alignment: $(v_1, v_2, \ldots, v_m) \in \mathscr{A}$. Gaps are included in the multiple alignment accordingly.

The quality of such a graph alignment will obviously depend on the choice of the pivot graph. Therefore, every graph is typically used as pivot structure and the best result is selected, provided that a quality measure for graph alignments is known. This adaption to the case of graph alignments has already been successfully used by Weskamp (2007).

Alternatively, other strategies from sequence alignments can be used, such as tree alignments (Böckenhauer and Bongartz, 2007), where multiple alignments are generated incrementally according to a guiding tree, e.g., realized in the form of the UPGMA algorithm in the ClustalW approach (Larkin et al., 2007). However, since the main purpose of the experimental validation in Chapter 5 is the comparison of the different algorithms, it should suffice to restrict to the pairwise case. This will make the interpretation of the results more comprehensible and keep the calculation times in a reasonable scale. Moreover, for some experiments, multiple alignments would not make much sense as will become apparent later on.

3.2 Derivation of graph models

The methods presented in this thesis are mainly designed for the structural comparison of protein binding sites, which are modeled in terms of graphs. Graphs represent a powerful and expressive data structure for the modeling of structured objects. Graph theory is a well-studied field in which many different algorithmic solutions to several graph-based problems have already been developed (Gross and Yellen, 2006).

The use of graph models for the purpose of modeling molecular structure data is especially appealing, since they offer a framework to represent the complete information inherently present in molecular structure data, without the need to invest prior knowledge. Of course, this could also be achieved otherwise, for example by using geometric models, which has also been done quite successfully (Fober et al., 2011; Shatsky et al., 2006).

However, another advantage of graph structures is the fact that they can transfer the coordinate-based information of geometric data into a relational form by using inter-component distances as edge weights. Thereby, each modeled entity is automatically set in the same frame of reference, superseding the need for translation and rotation prior to a comparison of the modeled objects.

In this thesis, graph representations of protein binding sites are based upon the CavBase database. The CavBase model represents a binding site model independent of sequence order or exact amino acid composition solely based on functional characteristics, physicochemical properties of amino acids that are capable of interacting with the corresponding ligands and substrates. This is achieved by introducing 3D descriptors of physico-chemical properties derived from surface-exposed amino acids.

The motivation behind this is to analyze protein-ligand interactions by focusing on the protein rather than the ligand itself and compare protein binding sites without utilizing ligand information. Since ligands can typically adopt multiple conformations, the question whether a specific protein can interact with a certain ligand is sometimes hard to answer when focusing on the ligand. Thus, this approach can be viewed as part of receptor-based drug design.

3.2.1 Extraction of protein binding sites

As outlined in Chapter 2, many strategies are known to extract protein binding sites. While the automated detection of protein binding sites is not a trivial task itself, the focus of this thesis is on the comparison of binding sites. All binding sites are thus derived from the CavBase database. CavBase uses the LigSite algorithm (Hendlich et al., 1997) to extract cavities on the surface of proteins. In LigSite, protein crystal structures derived from the PDB are embedded into a Cartesian grid with a 0.5 Å grid spacing. The surface of the protein is approximated by these grid points in the following way.

Initially, grid points are represented by 1.5 Å probe spheres and all spheres that intersect with the van der Waals radius of protein atoms are removed, since they are deemed solvent-inaccessible. For the remaining grid points the degree of burial b is estimated by counting the grid axes (the three Euclidean axes plus four cubic diagonals) that intersect with the protein structure. The more axes intersect, the higher the grid point is buried. Subsequently, grid points are clustered according to their degree of burial and clusters of highly buried grid points (with $b \geq 4$) are merged to form continuous cavities.

If no such cavities can be detected, the constraints are relaxed to search for more shallow cavities, more precisely, b is lowered. Each cluster of highly buried grid points must at least consist of 320 points, which roughly corresponds to a volume of 40 Å^3. This volume threshold guarantees that at least one water molecule can be accommodated.

The surface of a binding site is approximated by the grid points that are directly in contact with the protein atoms, non-buried grid points that are oriented toward the solvent are discarded. Additionally, cavity flanking amino acids are defined as residues, that are within a 1.1 Å range to a surface grid point. Surface grid points and cavity-flanking residues are subsequently stored in CavBase to represent information about the geometric shape of the binding sites. For the residues of the pocket, physicochemical descriptors are derived to model the binding site in terms of functional properties.

3.2.2 Modeling protein binding sites using physicochemical descriptors

The recognition of ligands and substrates is mediated through physicochemical interactions between the cavity-flanking residues and the ligand atoms, including hydrogen bonds, van der Waals interaction, metal/ion coordination and others (Bruno et al., 1997; McDonald and Thornton, 1994). Moreover, the solvent can mediate interactions between the binding site residues and ligand atoms and also covalent bonds are possible (Klebe, 2009).

On the one hand, one has to derive a model that is powerful enough to represent the interaction capabilities of the binding sites to make them amenable to algorithmic comparison and analysis. In other words, information about the spatial positioning and orientation of cavity flanking amino acids must be modeled to derive a meaningful representation for the inference of biochemical function. On the other hand, the model must be as compact and concise as possible to allow for an efficient algorithmic comparison.

As CavBase is built upon protein structure information derived from the PDB, much information about the protein binding sites is potentially available, including positioning of individual atoms. Obviously, the highest level of information would be achieved by deriving a full atom model of the binding site. However, in most cases a full atom representation of proteins is simply too complex to be of much use, especially in the context of database applications. This is also true for the comparison of protein binding sites.

A widely used alternative to model protein structure data is a C_α-representation, which common in structural bioinformatics (cf. Chapter 2). In this case, the spatial positioning of amino acids is represented by the coordinate of their C_α-atoms, thereby discarding any additional atom information. While this results in more compact models, such a representation seriously limits the information contained in the model: A standard C_α-representation of cavity-flanking residues would completely neglect the type of interactions these residues can participate in or at least obscure any information about the directionality of possible interactions or the conformational orientation and multiplicity of a side chain.

Thus, the CavBase model represents the geometric structure of the binding site with respect to the interactions the cavity-flanking residues can actually perform rather than a standard amino acid representation. As mentioned in Chapter 2, this is achieved by an abstraction of the pure structure data based on the cavity-flanking residues in terms of 3D descriptors, called *pseudocenters*.

Pseudocenters are three-dimensional physicochemical descriptors that represent certain types of molecular interaction an amino acid can participate in. Pseudocenters are assigned to certain groups of atoms of the cavity-flanking amino acids according to a fixed set of rules (Kuhn et al., 2006; Schmitt et al., 2002), thus creating a concise representation of binding sites in terms of their most important characteristics: geometric structure and the physicochemical properties.

The rationale behind this is the notion that the actual composition of an amino acid is not as relevant as the chemical property it provides. For example, if a ligand forms a hydrogen bond with a hydroxyl group of an amino acid side chain, it does not matter whether this group belongs to a tyrosine, serine or threonine side chain, as long as it is located at the right spatial position. Hence, a comparison based on these descriptors is theoretically more relevant.

Currently, CavBase distinguishes between seven types of pseudocenters:

1. hydrogen donor centers donate a polar proton to the formation of hydrogen bonds.

2. hydrogen acceptor centers represent corresponding acceptor positions for a hydrogen bond.

3. mixed donor/acceptor (doneptor) centers can contribute either a polar proton or an acceptor group for a hydrogen bond, for example in the case of hydroxyl groups.

4. pi centers represent the ability to form π-interactions perpendicular to the plane of aromatic rings and between carboxamide, carboxylate and guanidine groups.

5. aromatic centers have been introduced to account for the fact that aromatic rings can also form edge-to-face interactions (Kuhn et al., 2006). Hence, for aromatic residues a higher cutoff angle is considered compared to pi centers when filtering out pseudocenters with an unfavorable surface exposure (see below).

6. aliphatic centers represent the ability of non-polar side chain atoms (carbon and sulfur) to form hydrophobic interactions.

7. metal centers model the influence of coordinatively bound metal ions.

This model offers two benefits for the subsequent comparison of modeled binding sites. Firstly, representing the binding site geometry by more general physicochemical descriptors rather than the amino acids themselves, renders the representation much more tolerant towards mutations of the binding site if the positioning of important functional groups is conserved. Secondly, the model leads to a reduction of the number of 3D points that represent the cavity without abstracting too much, simplifying the algorithmic comparison of binding sites, since a smaller number of coordinates has to be considered. This is indeed an important issue as a comparison based on more complex representations (e.g., full atom representations) might be infeasible.

Fig. 3.2 summarizes the CavBase rules for assigning pseudocenters to protein binding sites as suggested by Schmitt et al. (2002) and later expanded by Kuhn et al. (2006). Subsequently, the pseudocenters are analyzed with respect to their surface exposure to remove pseudocenters that cannot possibly form an interaction with the ligand due to an unfavorable positioning. This is especially important for directional interactions such as hydrogen bonds.

This filtering is realized by measuring the angle between two vectors $\vec{v}$ and $\vec{r}$ as illustrated in Fig. 3.3. The vector $\vec{v}$ represents the mean orientation along which an interaction is most likely to be formed and $\vec{r}$ represents a normalized summation vector of all vectors from the pseudocenter to a neighboring surface point[2], respectively. The angle between these vectors is used as a cutoff angle to discard pseudocenters that are not likely to contribute to an interaction.

[2]A neighboring surface point is defined as surface point within a 3 Å radius of the pseudocenter

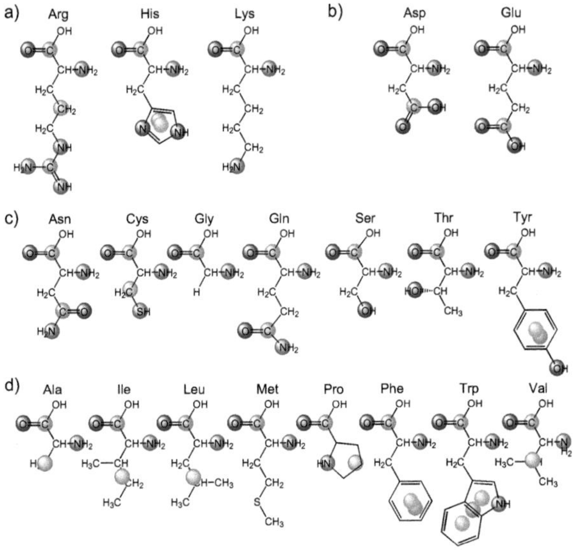

Figure 3.2: Summary of the assignment of pseudocenters according to the CavBase rules. Depicted are donor (red), acceptor (blue), mixed donor-acceptor (purple), pi (green), aromatic (cyan) and aliphatic (orange) pseudocenters. a) basic amino acids, b) acidic amino acids, c) polar uncharged amino acids, d) non-polar amino acids (for metal centers, no graphical example is displayed).

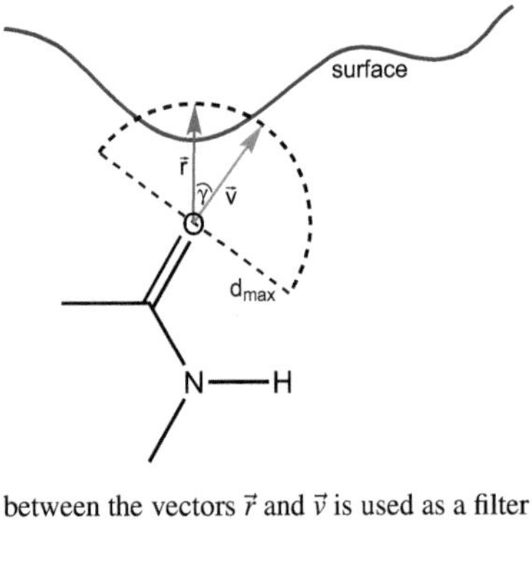

Figure 3.3: The angle between the vectors $\vec{r}$ and $\vec{v}$ is used as a filter criterion for pseudocenters.

A typical cavity as modeled by CavBase is depicted in Fig. 3.4. Note that this representation is independent of sequence order or fold information.

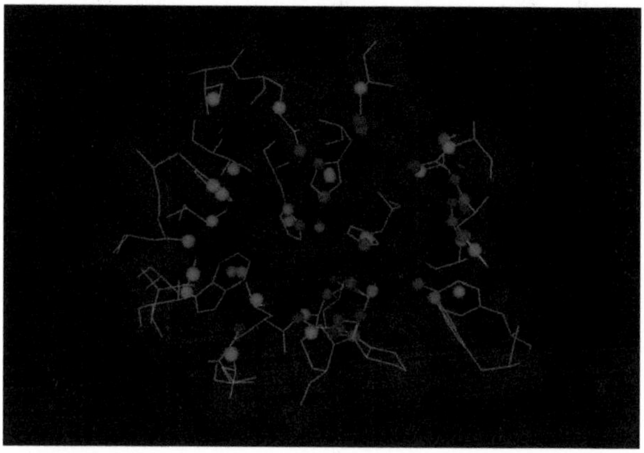

Figure 3.4: CavBase representation of a protein binding site. Bordering amino acids are shown in light blue, the semi-transparent surface indicates the Connolly surface. Pseudocenters are depicted as spheres (donor = red, acceptor = blue, donor/acceptor = purple, pi = gray, aromatic = green, aliphatic = cyan, metal = orange).

A set of pseudocenters constitutes an approximation of the spatial arrangement of physicochemical properties present in the binding pocket. As each pseudocenter is attributed to a point in three-dimensional space, it can be converted to a graph representation of the binding site by representing pseudocenters as nodes connected via edges labeled with the Euclidean distance between the pseudocenters.

A protein binding site is then modeled as a node-labeled and edge-weighted graph $G = \{V, E, l, w\}$ as defined above, with

- V denoting a set of nodes v corresponding to pseudocenters,

- $E \subseteq V \times V$ denoting a set of undirected edges connecting the nodes,

- $l : V \rightarrow \Sigma$, a label function, assigning the type of the corresponding pseudocenter to each node from the set of possible pseudocenter labels $\Sigma = \{$acceptor, donor,

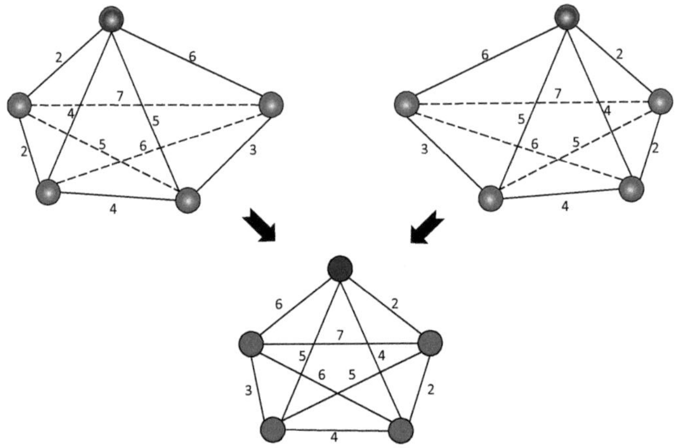

Figure 3.5: Two geometrically different constellations of pseudocenters, in fact mirror images of one another. Edge weights are depicted and node labels are represented by different colors. Note that it is not possible to transform one geometric structure into the other via transformation and rotation, hence these two bodies are not congruent. Yet, both would give rise to the same graph model.

doneptor, pi, aliphatic, aromatic, metal}

- $w : E \rightarrow \mathbb{R}_0^+$, a weighting function assigning the Euclidean distance between two pseudocenters to the edge connecting the corresponding nodes.

Note that the resulting graph model is a model independent of rotation and translation, superseding the need to find optimal coordinate transformations prior to the actual comparison, which would be necessary for geometrical models.

The theoretical downside of this representation is the fact that such a graph model is not an unambiguous geometrical model any more. In some cases it is possible for geometrically different constellations of points to give rise to identical graph representations, especially in case of mirror images of geometrical structures. An example is visualized in Fig. 3.5.

However, these cases are rather artificial compared to graphs derived from biological entities such as protein structures. Practically, as the graphs constructed from protein binding sites tend to be larger and more irregular in terms of geometry, it is assumed that this does not represent a serious limitation and in fact such graph models have already been applied successfully (Weskamp, 2007).

For the remainder of this work, the above defined graph model will be used to model protein binding sites derived from CavBase. It should be mentioned that in the original CavBase approach, the domain of edge weights was limited to a maximum weight of 12 Å to focus on low to mid-range distances, as inaccuracies increase for pseudocenters that are more distant (Schmitt et al., 2002). Moreover, pseudocenters in close proximity to each other are more likely to constitute a meaningful pattern, as interactions between the protein and a functional group of the ligand (e.g., an adenine moiety) will obviously occur in locally confined constellations. This constraint will be kept for the methods described in the next chapter with one exception. If an edge weight would exceeds this limit, the edge is omitted.

3.3 The CavBase approach and its implications

With the definition of the graph-theoretic concepts and the description of the graph models, it is now possible to take a closer look on the methods used so far in the CavBase scenario. As outlined in Chapter 2, the standard approach build into CavBase is a clique-based approach utilizing the well-known Bron-Kerbosch algorithm (Bron and Kerbosch, 1973).

More precisely, the clique approach is based on the assumption that two functionally related binding pockets accommodating the same binding partner will exhibit a similar spatial arrangement of properties, a similar motif that is necessary to establish the interaction with the binding partner.

This common motif can be detected by constructing a product graph as defined above and using the clique-enumeration approach of Bron and Kerbosch (1973) to detect cliques in the product graph. The maximum clique in the product graph then corresponds to the maximum common subgraph of both input graphs.

Given the specifics of the binding site model, the construction of the product graph has to be adjusted to account for the fact that mixed donor/acceptor pseudocenters can in principle provide an H-Donor group as well as an H-acceptor group. Hence, the matching criterion of two nodes must be weakened to allow for nodes with mixed donor/acceptor labels to be assigned to acceptor and donor pseudocenters as well.

Hence, the definition of the product graph is altered to account for these specifics. With $\Sigma = \{$acceptor, donor, doneptor, pi, aliphatic, aromatic, metal$\}$ and the node labeling function $l : V \to \Sigma$, the first condition in Definition 12 that a product node $(v_1, v_2) \in V_\times$ has to satisfy is changed from $l_1(v_1) = l_2(v_2)$ to

$$\begin{gathered} l_1(v_1) \in \{\text{acceptor, doneptor}\} \wedge l_2(v_2) \in \{\text{acceptor, doneptor}\} \vee \\ l_1(v_1) \in \{\text{donor, doneptor}\} \wedge l_2(v_2) \in \{\text{donor, doneptor}\} \vee \\ l_1(v_1) = l_2(v_2). \end{gathered}$$

Consequently, this relaxation will be kept for the approaches developed in this thesis.

The largest common motive of pseudocenter arrangements derived via clique-detection is of much interest for a functional analysis of the protein binding sites and can for example be recovered in the form of a graph alignment, in which the corresponding nodes of the common subgraph are mutually aligned. Moreover, for the purpose of a similarity search, the size of the MCS naturally gives rise to a similarity measure that can be exploited by similarity search procedures, for example in the context of a k-nearest neighbor search.

Let $G_1 = (V_1, E_1, l_1, w_1)$ and $G_2 = (V_2, E_2, l_2, w_2)$ be two graphs to be compared and $G_{MCS} = (V_{MCS}, E_{MCS}, l_{MCS}, w_{MCS})$ be the maximum common subgraph of G_1 and G_2, the similarity measure can be defined by

$$sim_{MCS}(G_1, G_2) = \frac{|V_{MCS}|}{\max(|V_1|, |V_2|)} \tag{3.2}$$

In the experimental part, this measure will serve as one of the baseline approaches for the validation.

The algorithm used internally in the CavBase database instead retrieves the largest 100 cliques using the Bron-Kerbosch algorithm and selects the best solution according to a surface-based scoring scheme[3]. This is done to exclude clique solutions with divergent corresponding surface regions[4] (Schmitt et al., 2002).

[3]This value was empirically determined to be the best compromise between efficiency and coverage of possible solutions (Schmitt et al., 2002)

[4]For example, cases can occur, where similar pseudocenter constellations are associated to either convex or concave surface patches.

3.4 Extreme value distributions

In the experimental section, the similarity of protein binding sites will be assessed using scores produced by the algorithm presented in the next chapter. Any score used to compare proteins or protein substructures must be judged against the likelihood that a given score could arise by chance. Therefore, it is necessary to obtain a measure of confidence. To this end, an empirical approach was chosen by calculating pairwise comparisons of randomly drawn cavities to derive a distribution of random scores and subsequently fitting a probability function to the obtained score distributions.

Significance for database searches can generally be assessed by extreme value distributions (EVD) (Stark and Russell, 2003) for the following reason: in the context of database searches, finding the most similar item to a query usually involves maximizing over all similarity scores between the query and the database items. Interpreting these scores as random variables, the maximum score can be viewed as an extreme value based on the score distribution. One can distinguish between three different types of EDV: The Gumbel family (type I)

$$f_{\mu,\sigma}(x) = \frac{1}{\sigma}\exp\left(\frac{x-\mu}{\sigma}\right)\exp\left(-\exp\left(-\frac{x-\mu}{\sigma}\right)\right) , \tag{3.3}$$

which is defined by the location parameter μ and the scale parameter σ, the (Fréchet) family (type II)

$$f_{\mu,\sigma,\xi}(x) = \frac{\xi}{\sigma}\left(\frac{\sigma}{x-\mu}\right)^{\xi+1}\exp\left(-\left(\frac{\sigma}{x-\mu}\right)^{\xi}\right) , \tag{3.4}$$

and the Weibull family (type III)

$$f_{\mu,\sigma,\xi}(x) = \frac{\xi}{\sigma}\left(\frac{x-\mu}{\sigma}\right)^{\xi-1}\exp\left(-\left(\frac{x-\mu}{\sigma}\right)^{\xi}\right) , \tag{3.5}$$

both of which have an additional scale parameter ξ.

Since the type of the EVD that should ideally be fitted to the score distributions obtained by the various methods was not known in advance, the scores were used to estimate the parameters of a generalized EVD

$$f_{\mu,\delta,\xi}(x) = \frac{1}{\sigma}\left[1+\xi\left(\frac{x-\mu}{\sigma}\right)\right]^{-1-\frac{1}{\xi}}\exp\left(-\left[1+\xi\left(\frac{x-\mu}{\sigma}\right)\right]^{-\frac{1}{\xi}}\right) \tag{3.6}$$

by maximum likelihood estimation. The GEVD transforms into the Gumbel distribution if $\xi = 0$. For $\xi > 0$ and $\xi < 0$, the GEVD yields the Fréchet family, respectively the Weibull family.

Subsequently, the corresponding cumulative distribution

$$F_{\mu,\sigma,\xi}(x) = \exp\left\{-\left[1+\xi\left(\frac{x-\mu}{\sigma}\right)\right]^{-\frac{1}{\xi}}\right\} \tag{3.7}$$

was used to calculate p-values for the comparison scores.

4

Methods

The comparison of protein binding sites has mainly been approached in a global manner using global comparison approaches (Binkowski and Joachimiak, 2008; Binkowski et al., 2003a; Kinoshita and Nakamura, 2005; Schmitt et al., 2002; Shulman-Peleg et al., 2004), which, from a methodological point of view, means that each approach tries to find a correspondence between protein binding sites by comparing them as a whole. Since protein binding sites can readily be modeled as graph structures, this translates into global graph comparison.

The main advantage of global graph-based approaches is the fact that, a mutual correspondences between components of the graphs can be derived by taking the whole graph topology into account. From this, a correspondence between basic structural units of the modeled structure can be established. The price for such a high yield of information is typically a high computational complexity, as finding an optimal correspondence comes down to solving a hard (combinatorial) optimization problem that is typically approached by means of heuristic methods. For example, the clique enumeration problem as well as the subgraph isomorphism problem are known to be NP-complete (Garey and Johnson, 1979; Karp, 2010). By extension, also the graph alignment problem is NP-complete, as it can be viewed as a generalization of the subgraph isomorphism problem (Weskamp, 2007).

Moreover, structure comparison in general has to deal with inaccuracies that inevitably arise when dealing with molecular data. These are due to measurement errors, crystallization artifacts, resolution issues, the uncertainty of amino acid side chain positions. Moreover, the dynamic flexibility of the molecules *in vivo* needs to be considered. This is especially a problem for global methods, which have to perform a balancing act between a certain degree of tolerance towards structural variation and the requirement to be specific and strict enough to recover meaningful similarities.

Thus, instead of focusing only on a global strategy, the idea of this work is to explore other options as well, capitalizing on the dualism between local and global concepts known from sequence comparison[1]. Given that protein binding sites are the entities to be compared, is a global comparison of protein binding sites more useful for the detection of shared biochemical or biological function or would it be advisable to use a local or semi-global strategy? All these principles certainly have merit.

In this chapter, several approaches to graph comparison are introduced utilizing conceptually different strategies, each one focusing on a different aspect of similarity. In Section 4.1, global graph comparison approaches are introduced, deriving a global graph alignment by comparing two or more graphs as a whole. Section 4.2 instead introduces approaches motivated from the field of kernel-based learning that derive a measure of similarity between graphs by comparing local substructures. Finally, both principles are combined to create a semi-global approach to graph comparison, which is presented in Section 4.3.

4.1 Global graph comparison

When modeling biological data, graph comparison has to account for structural variation, either on the model level, by defining an appropriate graph model, or on the methodological level, by developing comparison methods that exhibit a certain level of tolerance. In the context of CavBase, the latter alternative is considered by employing ε-isomorphism, as defined in the previous chapter, to tolerate certain edge weight differences resulting from inaccuracies on the protein structure level.

However, this cannot account for larger differences due to dynamic flexibility or a different arrangement of subpockets accommodating the same ligand in different conformation. Moreover, edge weight tolerance cannot account for mutations reflected by a change in node labels. A single mutation would not necessarily affect protein function, especially if the mutated pseudocenters do not play a major role in the interaction between protein and substrate or ligand. This is even more an issue for proteins that share a common function but lack a close hereditary relationship. Still, it would be desirable to

[1]Keep in mind that the terms "local" and "global" are used in a methodological sense, i.e., referring to the way in which graphs are compared. Thus, the correspondence to sequence-based methods is not exactly an analogy but rather an inspiring principle.

use a comparison method that is capable of detecting similarities despite such variance in order to uncover more remote similarities.

In the previous chapter, the modeling concept behind the CavBase database has been introduced. So far, the comparison of CavBase structures is realized using clique enumeration (Bron and Kerbosch, 1973) on an ε-tolerant product graph. While clique-detection offers a relatively fast approach to detect commonalities among protein binding sites, it also suffers from several drawbacks. Firstly, as it relies on the construction of a product graph, it suffers from a space complexity of $\mathcal{O}(m_1^2 m_2^2)$, with m_1 and m_2 denoting the number of nodes of two graphs. As some of the graphs derived from protein binding sites can reach a size of several hundred nodes, this renders the clique approach unable to calculate the MCS for larger input structures efficiently, simply due to memory limitations of current computers. This is especially problematic, since it has been shown that the active site of a protein is usually located in the largest cleft on the surface (Laskowski et al., 1996; Peters et al., 1996).

Another problem is the relative intolerance towards variation. By its very definition, the product graph discards nodes with different labels and varying coordinates are only tolerated as long as the distance relative to other centers is lower than ε. If pseudocenters deviate just a fraction of an Ångström beyond this threshold they will not be part of the solution. On the other hand, setting the ε tolerance too high obviously invokes the risk of becoming too unspecific.

Yet, as the main goal of this approach is the detection of a mutual correspondence between binding sites in the form of an alignment of pseudocenters, it might be advisable to allow for mismatches resulting from mutations in order to find an optimal overall correspondence, in analogy to mismatches in sequence alignments. In addition, given the different sources of inaccuracy when dealing with protein structures, tolerance towards structural deviation is mandatory.

The approach of Weskamp (2007) alleviates this second drawback by using the result of the clique-detection just as a "seed solution" to be extended by a greedy extension strategy. Still, the problem of space complexity remains, as this approach still depends on the solution of the clique-detection. Moreover, a greedy heuristic is a myopic optimization strategy, hence extending the clique solution greedily does not necessarily result in an optimal solution.

In the following, an alternative to the greedy strategy is presented that utilizes evolutionary optimization in order to alleviate the myopic nature of the greedy approach and sidesteps the requirement of a product graph, thereby making it more useful for the comparison of large binding pockets.

4.1.1 GAVEO - Global Graph Alignment Via Evolutionary Optimization

In the approach of Weskamp, the expansion of the seed solution is realized formulated as an optimization problem by using a graph edit distance as objective function. As outlined above, two graphs modeling the spatial topology of different binding sites cannot be expected to be isomorphic as some variation will inevitably exist even between closely related proteins.

To capture the difference between these graphs, a graph edit distance as mentioned in Chapter 3 can be used. The graph edit distance can be defined as the minimum number of edit operations necessary to convert a graph G_1 into another graph G_2 (Sanfeliu and Fu, 1983) (cf. chapter 2).

Here, three types of graph edit operations are considered to convert an arbitrary graph G_1 into another graph G_2:

1. Insertion: A node / edge is inserted into G_1.

2. Deletion: A node / edge is deleted from G_1.

3. Substitution: The label of a node or the weight of an edge in G_1 is altered.

By assigning different costs to the edit operations, Weskamp defined a scoring scheme that can be used to quantify the difference between graphs and fulfills the properties of a metric (Weskamp, 2007). To model the deletion of nodes, Weskamp introduced so-called *dummy nodes*, placeholders for the deleted nodes which will correspond to gaps in the context of graph alignments. The resulting function represents a quality measure for alignments and is used as objective function to be maximized by a greedy heuristic that expands upon the seed solution. A multiple graph alignment is subsequently constructed by employing the star alignment merging technique described in Chapter 3.

The Weskamp approach still suffers from several drawbacks. Firstly, the problem of space complexity remains, since a vital part of the approach consists of finding the maximal clique in a product graph. Secondly, greedy heuristics in general suffer from the drawback of being short-sighted. So-called myopic optimization strategies carry a high risk of getting stuck in a local optimum. This is due to the fact that backtracking is impossible, hence a decision once made cannot be undone, even if it would lead to a better solution. Likewise, look-ahead strategies are not employed.

In the following, an alternative algorithm based on evolutionary optimization called GAVEO (Global Alignment Via Evolutionary Optimization) is introduced that circumvents both problems. Evolutionary algorithms represent a number of metaheuristic optimization strategies inspired by the biological concept of evolution. Realizing a (non-myopic) EA, the GAVEO algorithm offers the potential to yield a significant improvement in terms of alignment quality by avoiding local optima. Moreover, GAVEO eliminates the need to generate seed solutions. This effectively allows to omit the clique-detection step, thereby circumventing the space complexity problem.

4.1.1.1 An evolutionary strategy for the calculation of graph alignments

As outlined above, the problem of finding an optimal graph alignment can be formulated as an optimization problem. An evolutionary algorithm (EA) offers the benefit that each point in the search space can be reached which allows for a more thorough exploration of the search space. Moreover, evolutionary algorithms have proven to be relatively versatile, leading to (near-)optimal solutions for a variety of different optimization problems (Spears et al., 1993). However, while EAs in principle are capable of finding an optimal solution in finite time, they cannot guarantee to find it in a reasonable amount of time and in fact high runtime requirements are the major downsides of EAs (Ashlock, 2006).

Evolutionary algorithms use computational models of evolutionary processes inspired by Nature to solve optimization problems. As a result, the terminology of evolutionary computation draws heavily on terms used in biology in the context of evolution. EAs typically maintain a set of possible candidate solutions for a given problem called *individuals* that are iteratively refined by applying a reproduction and selection regime (Bäck et al., 1997, 2000; Spears et al., 1993). In each iteration, individuals are perturbed by applying search operators typically referred to as *mutation* and *recombination* that serve as exploration heuristics to explore the search space. Subsequently, individuals are subjected to *selection*, by evaluating the perceived performance of the individuals measured by a *fitness* function and selecting certain individuals according to a specified selection scheme as *offspring* for the next iteration. The set of individuals is referred to as *population* and an iteration, in accordance with the evolution symbolism, is also called *generation*.

The GAVEO algorithm builds upon the framework of (Weskamp, 2007). More precisely, GAVEO calculates a global graph alignment as defined in Chapter 3 by maximizing a global scoring function s that serves as a fitness function. However, instead of starting from precalculated seed solutions, GAVEO offers the possibility to calculate graph alignments "from scratch" starting from randomly generated alignments.

Given a set of graphs $\mathbb{G} = \{G_1, ..., G_m\}$, a graph alignment $\mathscr{A}$ is calculated that maximizes the optimization function $s(\mathscr{A})$. The objective function used by GAVEO is identical to the one used by Weskamp and represents a quality measure based on a sum-of-pairs scheme, generalized for the case of multiple graph alignments. The score of a multiple alignment $\mathscr{A} = (a^1, \ldots, a^n)$ is calculated by summing over the scores of all induced pairwise alignments:

$$s(\mathscr{A}) = \sum_{i=1}^{n} \mathrm{ns}(a^i) + \sum_{1 \leq i < j \leq n} \mathrm{es}(a^i, a^j) \ . \tag{4.1}$$

The function consists of two parts, considering nodes and edges separately. The node score ns evaluates the correspondence of all mutually assigned nodes within a column a^i of the alignment $\mathscr{A}$, which is summed up over the length of the alignment. Matching node labels are rewarded by a positive score ns_m, mismatches or the assignment of gaps are penalized by negative values ns_{mm} and ns_{gap}, respectively:

$$\mathrm{ns}\begin{pmatrix} a_1^i \\ \vdots \\ a_m^i \end{pmatrix} = \sum_{1 \leq j < k \leq m} \begin{cases} \mathrm{ns}_m & \ell(a_j^i) = \ell(a_k^i) \\ \mathrm{ns}_{mm} & \ell(a_j^i) \neq \ell(a_k^i) \\ \mathrm{ns}_{gap} & a_j^i = \perp, a_k^i \neq \perp \\ \mathrm{ns}_{gap} & a_j^i \neq \perp, a_k^i = \perp \end{cases} \tag{4.2}$$

The function es evaluates the assignment of edges. Tolerance towards edge weights deviation is again realized by ε tolerance. Thus, the assignment of two edges e_1 and e_2 is considered a match, if the respective weights deviate by ε at most, otherwise mismatch is presumed:

$$\mathrm{es}\left(\begin{pmatrix} a_1^i \\ \vdots \\ a_m^i \end{pmatrix}, \begin{pmatrix} a_1^j \\ \vdots \\ a_m^j \end{pmatrix}\right) = \sum_{1 \leq k < l \leq m} \begin{cases} \mathrm{es}_{mm} & (a_k^i, a_k^j) \in E_k, \ (a_l^i, a_l^j) \notin E_l \\ \mathrm{es}_{mm} & (a_k^i, a_k^j) \notin E_k, \ (a_l^i, a_l^j) \in E_l \\ \mathrm{es}_m & d_{kl}^{ij} \leq \varepsilon \\ \mathrm{es}_{mm} & d_{kl}^{ij} > \varepsilon \end{cases} , \tag{4.3}$$

where $d_{kl}^{ij} = \left| w(a_k^i, a_k^j) - w(a_l^i, a_l^j) \right|$. The parameters (i.e., ns_m, ns_{mm}, ns_{dummy}, es_m, es_{mm})

are constants used to reward or penalize matches, mismatches and dummies, respectively[2].

Having defined the fitness function, an *evolutionary algorithm* is employed to find a globally optimal alignment. More precisely, the GAVEO approach consists of an iterative process according to Beyer and Schwefel (2002). Initially, a population consisting of μ individuals is generated randomly, with μ denoting the population size. Each individual represents a graph alignment as candidate solution. Then, the following iterative loop is performed until a certain stopping criterion is met:

1. At the beginning of each generation, $\lambda = \nu \cdot \mu$ new offspring individuals are generated. This is achieved by selecting ρ parent individuals from the initial population by means of a mating-selection operator, which in the case of GAVEO is simply realized as a random selection according to a uniform distribution. The parent individuals are then *recombined* to yield a new individual. This is realized by means of a recombination operator.

2. Each individual is subjected to a mutation operator, that (slightly) alters the individual.

3. The offspring individuals are subsequently evaluated using the fitness function 4.1 and a temporary population T is formed as union of the initial population and the newly generated offspring. A selection operator is then applied to select the fittest μ individuals that form the population off the next iteration.

As possible stopping criteria, the elapsed runtime, the number of generations, stall time or stall generations (the amount of time, respectively the number of generations, with no improvement of the fitness value) or a fixed fitness value could be used. The complete GAVEO algorithm is summarized in Algorithm 1.

The optimization process can start from arbitrary alignments which allows to reduce the memory requirements compared to the greedy approach, as mentioned above. However, since the search space is relatively large, starting from a random alignment might not be the best choice. Thus, in case of smaller graphs where the calculation of a maximum clique is unlikely to cause problems, the greedy solution can be calculated as starting point first and included in the initial population.

[2]In the experimental part, the scoring parameters and ε will be initialized accoring to Weskamp (2007).

Algorithm 1 The GAVEO algorithm

Require: $\mathbb{G}$ set of graphs, μ population size, λ number of offspring, ρ number of parents, s fitness function
  $stop = false$
  P = initialize_population($\mathcal{G}, \mu$)
  while stop = false **do**
    O = recombine_offspring(P, λ)
    O = mutate_offspring(O)
    evaluate_population($O \cup P$)
    P = select_best($O \cup P, \mu$)
    if stop criterion is met **then**
      $stop = true$
  return individual $A \in P$, with $A = \arg\max_{A \in P} s(A)$

In principle, any number of graphs can be aligned in this manner directly, which is an additional benefit of the GAVEO approach. The greedy approach instead can only be used to calculate pairwise alignments that are subsequently combined into a multiple graph alignment via star alignment. This, however, introduces another source of inaccuracy, as two heuristics are used in combination instead of one single procedure, as is the case for the GAVEO approach.

One the other hand, the search space of the multiple graph alignment problem grows exponentially with the number of graphs. This, is of course problematic from an optimization point of view, as an efficient exploration of the search space becomes more and more difficult. Thus, decomposing the multiple graph alignment problem into several pairwise ones and resorting to subsequent merging might allow to trade quality for speed.

In doing so, one would achieve a reduction of the search space by simplifying the problem, although bought with the potential loss of quality incurred by the sub-optimal merging of the pairwise alignments. It is difficult to judge which effect would be of greater impact in advance and it might be more advisable to avoid the risk of getting astray in a huge search space. However, as the focus of this thesis is on the pairwise case, this is of minor importance here.

4.1.1.2 Initialization and representation of individuals

In many cases, evolutionary algorithms, especially evolutionary strategies (Rechenberg and Eigen, 1973), utilize a problem-specific representation of individuals. Here, a graph alignment is represented by using a $m \times l$ matrix that holds the indices of mutually assigned nodes, with m denoting the number of graphs to be aligned and l the length of the alignment. More precisely, each row corresponds to a graph and each column to a set of mutually aligned nodes. By introducing an arbitrary but fixed numbering of nodes for each graph, every number in the matrix uniquely identifies a certain node in a graph. An example is given in Fig. 4.1. As edges are specified indirectly as well, this representation uniquely encodes a multiple graph alignment.

G_1:	1	2	-	4	3	5	-
G_2:	3	2	6	1	4	5	-
G_3:	4	3	2	-	1	-	-
G_4:	-	3	4	-	2	1	-

Figure 4.1: Matrix representation of an MGA. The first column indicates a mutual assignment of the first node of graph 1, the third node of graph 2, and the fourth node of graph 3, while there is no matching partner in graph 4. Gaps are represented by -. Note that the order of the columns is arbitrary.

The question arises how the length of the alignment is set. Since the definition of a graph alignment as introduced above only requires that each column must at least contain one non-gap position, the length of the optimal alignment $\mathscr{A}$ can be upper-bounded by $|\mathscr{A}| \geq \sum_{i=1}^{m} |V_i|$ but is generally not known in advance. For practical reasons, setting the second dimension of the matrix to this upper bound is not advisable, as this will again result in high storage requirements that might cause problems similar to the product graph calculation.

Moreover, from an optimization point of view, the potentially huge search space would make the search process unnecessary excessive under the assumption that the optimal solution will almost never occupy the whole length. While for efficiency a smaller alignment length is preferable, setting a lower limit bears the risk of excluding the optimal solution from the set of possible solutions that can be obtained.

To solve this dilemma, an adaptive representation is used. For each individual, a single gap column is added to the solution serving as a reservoir of placeholders to be matched where appropriate. The idea is that, if inserting a gap in the alignment would be favorable, the gap column will be disrupted by placing one or more placeholders at a new position within the alignment in exchange with the node that was previously assigned. Further gap columns will be introduced automatically if that happens, or, in the opposite case, will be removed if they accumulate. This ensures that always one such column is present.

During optimization, it is ensured that gap columns have no influence on the fitness of the individual.

Obviously, this will not result in a limitation of the search space, as the maximum length alignment can still be achieved. However, this strategy creates a bias towards shorter solutions which should improve the runtime requirements of GAVEO. As a result, the second matrix dimension will be set to $l = \max\{|V_1|, \ldots, |V_m|\} + 1$ to accommodate the largest graph and the additional gap column. During initialization, μ individuals are generated. For each graph $G_i \in \mathbb{G}$, a permutation of the nodes in $|V_i|$ is inserted into the i-th row of the matrix and for each index position $j > |V_i|$ a placeholder is inserted.

4.1.1.3 Evolutionary operators

As GAVEO uses a problem-specific representation, standard evolutionary operators known from the field of evolutionary optimization do not apply. Hence, these operators have to be adapted for GAVEO. In the following, a concrete description of the operators is given.

Recombination

The recombination operator constructs a new individual from ρ parent individuals drawn at random from the current population via a uniform distribution. To select the submatrices to be combined, $\rho - 1$ random numbers r_i, with $i = 1, \ldots, \rho - 1$ are generated with $1 \leq r_1 < r_2 < \ldots < r_{\rho-1} < m$, specifying the rows to be taken from the individual parents. To obtain the new offspring, the rows $\{r_{i-1} + 1, \ldots, r_i\}$ from the i-th parent individual are selected (where $r_0 = 0$ and $r_\rho = m$ by definition) and combined.

As the indices in a row are not ordered, simply concatenating the rows as they are is not reasonable, since this would disrupt the reference frame of the row. This would be counterintuitive, as it violates the idea behind the recombination step, which is to combine elements from the parent individuals in order to allow the combination of (hopefully) favorable assignments from the parent individuals in the offspring.

Therefore, in a merging step i, the ordering of the r_i-th row is used as a pivot row, a reference in order to preserve assignments already present in the parents. The submatrices derived from the parent individuals as specified by the rows $\{r_{i-1} + 1, \ldots, r_i\}$ are combined columnwise, by joining each subcolumn with the subcolumn of the next individual that has the same node index entry in the pivot row r_i. In case the entry in row r_i is a gap, which can occur multiple times, the first occurrence of a gap is chosen from the next individual, marking this column as "used".

This procedure is illustrated in Fig. 4.2 for the case $\rho = 3$. Three individuals I_1, I_2, and I_3 and two rows designated by the integers r_1 and r_2 with $1 \leq r_i \leq m$ are chosen at random. All individuals are split horizontally at the rows r_1 and r_2. The resulting blocks are then merged into a new offspring individual. To preserve the ordering of the parent individuals, columns are rearranged according to the reference rows r_1 and r_2, respectively, whose indexes serve as pivot elements. In the illustration, the first red subcolumn in I_1 is transferred to the offspring, with the index of the pivot row r_1 being 2. The column is then expanded by searching for the occurrence of the index 2 in the pivot row r_1 in the next individual and transferring the associated next subcolumn (red) to the offspring individual. This procedure is repeated for all individuals and columns.

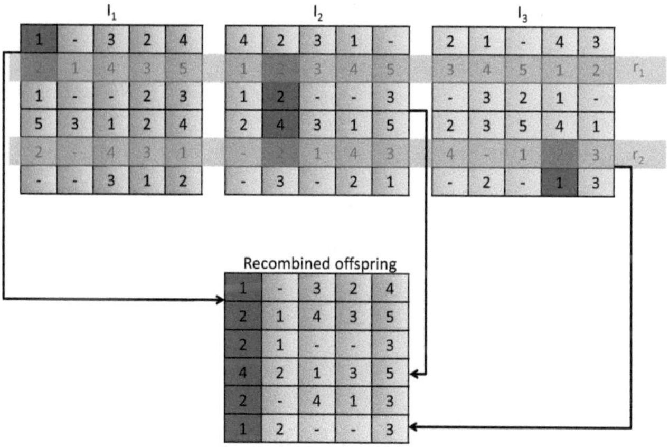

Figure 4.2: Recombination of $\rho = 3$ individuals. r_1 and r_2 designate the pivot rows (green), where the parent individuals are split. The red subcolumns are combined in a new offspring individual, preserving the assignment of nodes from the parent individuals.

Mutation

The performance of an EA largely depends on the mutation operator, which is guided by two contrary principles (Beyer and Schwefel, 2002). On one hand, it needs to allow for minimal changes to ensure that the exploration of the search space is fine enough to be able to reach every point, that is, every possible alignment. Moreover, if a near-optimal solution has already been reached, this ensures that one mutation step will not deviate much from this solution. On the other hand, larger steps are necessary to avoid premature stagnation and allow for a more rapid exploration of the search space. This trade-off is controlled by the mutation strength.

In GAVEO, the mutation operator is realized in a relative simple way, by randomly selecting a single row r and swapping two randomly chosen entries r_i and r_j. The mutation strength is regulated by performing this mutation steps repetitively, with the number of repetitions corresponding to the mutation strength. Fig. 4.3 illustrates the mutation operator.

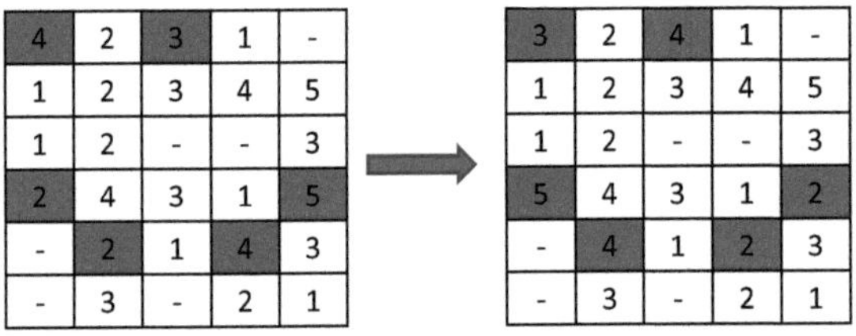

Figure 4.3: Mutation of an individual with a mutation strength of 3.

This way, an adjustable mutation operator is created that allows for the determination of the most successful mutation strength by specifying the mutation strength as a strategy component that can be adjusted instantly using a self-adaptation mechanism (Beyer and Schwefel, 2002). This is necessary, since the optimal mutation strength is not known in advance.

Adaption of Alignment Length

The adaptation of the alignment length, as described above, occurs at randomly chosen intervals. For a given individual, the algorithm checks with a random probability p_{check} whether an extension of the alignment is necessary. To this end, the presence of a gap column is checked. Three cases can occur:

1. Exactly one gap column is present. The alignment length does not have to be adjusted, as there are still placeholders available for every row.

2. Gap columns have accumulated, indicating that a matching of nodes is more favorable than introducing gaps, which in turn are considered obsolete. The number of gap columns is reduced to one.

3. No gap column is present, indicating that gaps have been introduced in the alignment to improve alignment quality. To restore the reservoir of placeholders, a new gap column is inserted.

Selection

As a selection operator, the deterministic plus-selection was chosen, thus realizing a $(\mu+\lambda)$ scheme known from evolutionary strategies (Beyer and Schwefel, 2002). In a $(\mu+\lambda)$ strategy, the μ individuals of the parental generation are chosen to create additional λ offspring individuals. During the selection process, all individuals are evaluated according to the fitness function 4.1. The population of the next generation is then created by selecting the best μ individuals according to their fitness.

For the current problem, this is arguably the most promising strategy, given that the search space for the multiple graph alignment problem is extremely large with a size of $\mathcal{O}(k!^{m-1})$ (k denoting the length of the alignment, which is a priori not known and m the number of graphs). Thus, it would be advisable to regard the parent individuals as well as the offspring to ensure that no currently best solution is lost. An (μ,λ) strategy for example, utilizing the comma-selection (Beyer and Schwefel, 2002), which only considers the offspring, would discard the parent generation, regardless of their fitness value.

4.1.1.4 GAVEOc - keeping the clique solution

As mentioned above, calculating a new graph alignment starting from a random initialization allows for solving the space complexity problem, but on the other hand bears the risk of increasing the runtime. Therefore, the possibility to use the greedy solution as a starting point was provided, which apparently also includes the MCS as derived by clique enumeration. Obviously, the trade-off here is the fact that a complete greedy solution must be calculated in advance, before the real optimization starts.

However, since the greedy solution included in the start population is subsequently subjected to another optimization, one could also skip this step entirely. Calculating the MCS instead is much more interesting, as the largest common substructure of the binding pockets might contain important information that should not be lost. In fact, a large common substructure in the pocket would be highly meaningful. While the main problem of using the MCS as similarity criterion is its lack of flexibility, there is no reason to dismiss this information if available, thus it should be part of the final solution.

Therefore, as an alternative, the above algorithm is altered in the following way: Prior to the optimization process, the MCS is calculated once via the Bron-Kerbosch method and included in *every* individual of the starting population, instead of only one individual. Moreover, the MCS is *kept* during the optimization process.

The GAVEO approach allows for a complete exploration of the search space. That means, theoretically, even if a large common subgraph would exist, it could be disrupted by the optimization process, if that would lead to an increase of the objective function. The fitness function mainly operates on the assumption, that a common substructure, even if containing some mismatches, would still yield an optimal score if matched accordingly. However, since the sum-of-pairs score considers all assignments independently, this is not necessarily the case.

Thus, to ensure the intactness of the MCS, the largest obtained clique solution is fixed and the optimization is limited to the remainder of the alignment. To this end, the clique solution is set as a "prefix" of the complete alignment, and the mutation operator of GAVEOc selects row numbers randomly from the interval $[0+c, l]$, c denoting the size of the MCS. While this GAVEOc variant might not necessarily lead to a globally optimal alignment in terms of the above-defined scoring function, it might well lead to better results in a biological sense.

4.1.1.5 Alignment scores

As outlined above, the scoring function can be regarded as a similarity measure for the aligned structures: The higher the number of matches, the higher the score and the more similar the binding pockets. However, using the fitness value directly as a similarity measure might be ill-advised.

Hypothetically, consider the case of three binding pockets that all have the same common core structure and a different number of additional pseudocenters at the rim of the cavity. In case that the additional nodes and edges can only be assigned as mismatches, the two smallest graphs, respectively binding pockets, would receive the highest similarity score. Yet, obviously, it would be more reasonable to score all three graphs as equally similar, especially since the extraction of cavities is inaccurate at the borders.

It would be even worse if binding pockets with a smaller common core structure might receive a higher similarity score than pockets with a larger one, provided that the number of non-matching nodes and edges in the latter case is large enough to decrease the score below the value of the first comparison. This is depicted in Fig. 4.4. An alignment $\mathscr{A}_{12}$ between the graphs 1 and 2 would yield a similarity score of $s(\mathscr{A}_{12}) = (2 \cdot \mathrm{ns}_m + 1 \cdot \mathrm{es}_m) - (5 \cdot \mathrm{ns}_{gap} + 7 \cdot \mathrm{es}_{mm})$, whereas an alignment between graphs 2 and 3 would be scored $s(\mathscr{A}_{12}) = (4 \cdot \mathrm{ns}_m + 5 \cdot \mathrm{es}_m) - (8 \cdot \mathrm{ns}_{gap} + 17 \cdot \mathrm{es}_{mm})$. Assuming, that $\mathrm{ns}_m = 1$, $\mathrm{ns}_{gap} = -1$,

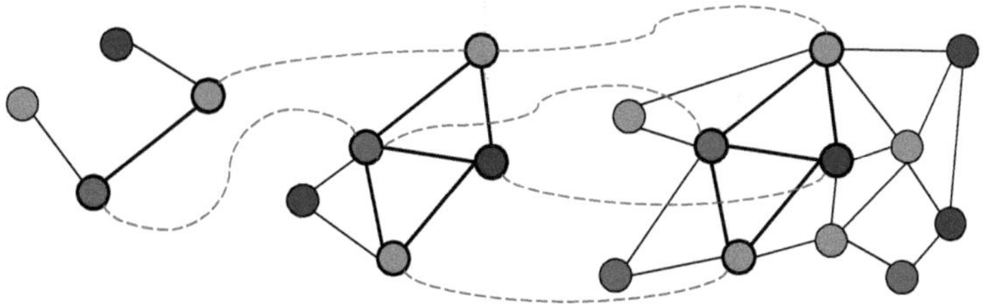

Figure 4.4: Example of a counterintuitive similarity degree based on the objective function of GAVEO. Blue dotted lines indicate the alignment of nodes, thick lines indicate common subgraphs.

$\mathrm{es}_m = 0.1$ and $\mathrm{es}_{mm} = -0.1$, the former alignment would be scored higher ($s(\mathscr{A}_{12}) = -3.5$) as the latter ($s(\mathscr{A}_{12}) = -5.2$), despite the fact, that graphs 2 and 3 have a larger common subgraph.

Thus, alternatively, the following similarity score will be used:

$$s'(\mathscr{A}) = \sum_{i=1}^{n} \mathrm{ns'}(a^i) + \sum_{1 \le i < j \le n} \mathrm{es'}(a^i, a^j) \ , \tag{4.4}$$

with

$$\mathrm{ns'}\begin{pmatrix} a_1^i \\ \vdots \\ a_m^i \end{pmatrix} = \sum_{1 \le j < k \le m} \begin{cases} \mathrm{ns}_m & \ell(a_j^i) = \ell(a_k^i) \\ 0 & \text{otherwise} \end{cases} \tag{4.5}$$

$$\mathrm{es'}\left(\begin{pmatrix} a_1^i \\ \vdots \\ a_m^i \end{pmatrix}, \begin{pmatrix} a_1^j \\ \vdots \\ a_m^j \end{pmatrix}\right) = \sum_{1 \le k < l \le m} \begin{cases} \mathrm{es}_m & d_{kl}^{ij} \le \varepsilon \\ 0 & \text{otherwise} \end{cases} \tag{4.6}$$

The problem still remains that this similarity score is not normalized, i.e., it will depend on the size of the graphs. To obtain a size-independent similarity measure, the above defined score is normalized to the size of the graphs. This gives rise to two similarity measures, a "conjunctive" and a "disjunctive" one:

$$\zeta_{max}(\mathscr{A}) = \max\{\frac{s'(\mathscr{A})}{|G_1|}, \frac{s'(\mathscr{A})}{|G_2|}\}, \quad (4.7)$$

$$\zeta_{min}(\mathscr{A}) = \min\{\frac{s'(\mathscr{A})}{|G_1|}, \frac{s'(\mathscr{A})}{|G_2|}\}. \quad (4.8)$$

The measure (4.8) can be seen as a relaxed equality and proceeds from the expression of set equality ($A = B$) in terms of two-sided inclusion ($A \subset B$ and $B \subset A$). Thus, it requires that, to be similar, G_1 and G_2 must be approximately equal in the sense of a mutual inclusion: G_1 is (approximately) included in G_2 and likewise G_2 in G_1. As opposed to this conjunctive combination of the two degrees of inclusion, the disjunctive combination (4.7) only requires an inclusion on one side: either G_1 is included in G_2 or G_2 in G_1. Obviously,

$$\zeta_{min}(G_1, G_2) \leq \zeta_{max}(G_1, G_2) \ .$$

Both measures can be regarded as complementary. The question which of these two measures, the conjunctive or the disjunctive one, yields more suitable similarity degrees cannot be answered in general and instead depends on the problem setting, in particular on the purpose for which the similarity is used (e.g., function prediction) and the way in which protein binding sites are extracted and modeled (e.g., whether or not the model may include parts of the protein not belonging to the binding site itself).

Therefore, to allow for a certain flexibility and account for both possibilities, a combination of both will be used as ultimate similarity measure:

$$\zeta(G_1, G_2) = \alpha \cdot \zeta_{max}(\mathscr{A}) + (1 - \alpha) \cdot \zeta_{min}(\mathscr{A}) \ . \quad (4.9)$$

As a remark, it should be noted that, formally, equation (4.9) is a special case of a so-called OWA (ordered weighted average) aggregation (Yager, 1988), with the parameter $\alpha \in [0, 1]$ controlling the trade-off between the two similarity measures.

4.2 Local graph comparison

In the previous section, the problem of graph comparison was addressed in a global way. As mentioned above, taking the whole graph topology into account to derive a mutual correspondence between binding sites comes at the price of a high computational complexity. Moreover, the approaches discussed in the previous chapter inherently bear the risk of having relatively high runtime requirements, as evolutionary optimization is known to be expensive (Ashlock, 2006). As a result, global graph comparison might not prove efficient enough to be used in the context of large-scale *in silico* screening, for example, when screening the CavBase for similar binding pockets, given a query protein binding site that represents a new drug target to detect cross-reactivities.

With these considerations in mind, one might raise the question whether it is really necessary to derive a global alignment of protein binding sites in a relatively complex manner, or whether the same task can be performed by much faster and simpler, although maybe less powerful methods with respect to the information obtained. For the purpose of database searches, for example, a complete assignment of nodes (respectively pseudocenters) is not really necessary. In fact, the derivation of a similarity measure would suffice for this type of application.

But the comparison of binding sites in a global manner can also be questioned from a biological point of view. Firstly, protein structures as such are flexible, subjected to conformational changes and dynamic behavior. Global approaches considering the whole graph topology are more easily affected by disturbances in the topology, which can easily occur as a result of conformational changes and inaccuracies of the modeled structures. While the issue of flexibility has been approached by several algorithms to protein comparison (Shatsky et al., 2004; Verbitsky et al., 1999), little has been done in the field of graph methods beyond the use of tolerance thresholds.

As a result, global methods might fail to detect similarities among related proteins in different conformations. A simple illustration of this effect is shown in Fig. 4.5. The two "geometric" graphs depicted are almost identical, except for the variation of the angle at the red node. Yet, this small modification already affects the whole graph topology, for example the length of the edges indicated by dashed lines.

Secondly, functionally related structures may only share similar substructures, a notion that has been supported by recent studies (Najmanovich et al., 2008). This is also true for protein binding sites. As cleft-detection algorithms (e.g., based on alpha-shapes or grid scanning) in general suffer from an inaccuracy in determining the borders of the cavity, different binding site representations can be derived even for the same protein. Moreover, ligands might occupy only a small portion of a cavity, hence functionally related binding pockets do not necessarily share the same overall architecture.

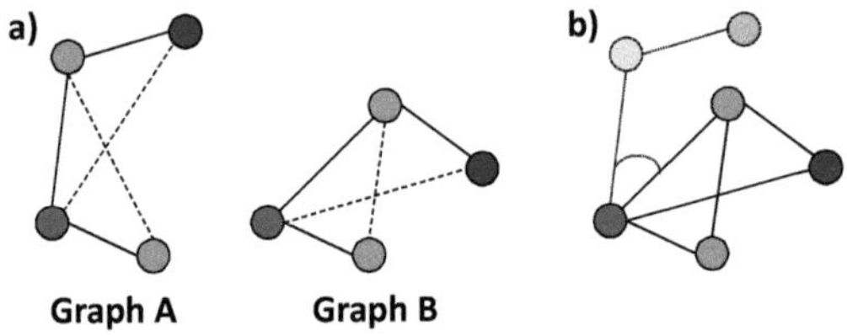

Figure 4.5: Two almost identical graphs (a), except for the variation of the angle at the red node, which influences the length of several edges (dashed). An overlay of the two graphs (b) shows a change in graph topology.

While it would be desirable to accurately determine the actual part of the binding pocket that interacts with the ligand in order to exclude irrelevant information, this is not feasible based on the protein structure alone, given the fact that ligands usually possess many rotatable bonds leading to a high number of degrees of freedom. Furthermore, using additional ligand information is generally not viable, since for many proteins such exhaustive information about all possible interacting ligands is simply not available.

But beside these technical issues, even if the protein binding pockets are reduced to the relevant interacting part, a global similarity between binding sites interacting with the same ligand must still not necessarily exist and, by extension, no overall identical graph topology. As a result of the freely rotatable bonds within a ligand, the ligand can adopt different conformations. Thus it is plausible that subpockets of a binding site interacting with certain functional groups of a ligand can be arranged differently and still accommodate the same ligand, although in a different conformation.

Therefore, an obvious alternative would be to approach the problem of graph comparison on a local scale, thus transferring the graph similarity problem to the level of substructures. This way, a similarity measure can be derived by comparing local graph features, e.g., subgraphs of a specific type. A high degree of similarity is already obtained by two graphs having similar constituents instead of an isomorphic global topology. Conceptually, this can be viewed as orthogonal to global graph comparison.

Significant contributions in this direction have been made in the form of so-called *graph kernels*. Roughly speaking, a kernel can be regarded as a similarity function on arbitrary complex objects, in this case graphs, fulfilling certain properties. Mathematically, a kernel is defined as follows:

Definition 16 *(Positive definite kernel)*

Let $\mathcal{G}$ be a set of objects. A function $\kappa : \mathcal{G} \times \mathcal{G} \to \mathbb{R}$ is called kernel if the following properties hold:

1. $\kappa(x,y) = \kappa(y,x)$ *for all* $x,y \in \mathcal{G}$ *(symmetry),*

2. $\sum_{i,j=1}^{m} c_i c_j \kappa(x_i,x_j) \geq 0$, $\forall m \in \mathbb{N}, \forall \{c_1,\ldots,c_m\} \subseteq \mathbb{R}$ *and* $\{x_1,\ldots,x_m\} \subseteq \mathcal{G}$ *(positive definiteness).*

From a machine learning point of view, kernels are especially attractive and interesting tools, as they offer the possibility to employ linear classifiers to solve non-linear problems. More precisely, every linear classifier that solely depends on a dot product of the instances can be employed in such a manner, the most prominent example being support vector machines (SVMs).

SVMs implicitly utilize the concept of a mapping function $\Phi : X \to \mathcal{H}$ to map instances not linearly separable in input space into a higher-dimensional feature space, the so-called *Hilbert space* $\mathcal{H}$, where the data becomes linearly separable. Conveniently, the mapping does not have to be explicitly calculated. Instead, a kernel can be plugged into the equation, since the SVMs solely depend on the dot product of the mappings. This is known as the so-called *kernel trick* (Aizerman et al., 1964). Interestingly, kernels cannot only be defined on feature representations but in principle on arbitrary complex objects, such as graphs. By defining a suitable kernel function $\kappa(G_1,G_2) = \langle\Phi(G_1),\Phi(G_2)\rangle$, graphs become amenable to kernel-based machine learning.

It should be noted that it is also possible to convert the kernel measure into a distance metric, the so-called *kernel distance*, via

$$\delta(x,y) = \sqrt{\kappa(x,x) - 2\kappa(x,y) + \kappa(y,y)} \ . \tag{4.10}$$

However, since graph kernels can also be regarded as similarity functions on graphs, this conversion is not really necessary, since they can also be used directly, for example, in the context of similarity retrieval and clustering tasks.

4.2.1 Extension of existing R-convolution kernels

The question is how to derive a kernel function on structured objects in general and graphs in particular. Intuitively, when defining a similarity function on arbitrary structured objects, one could decompose them into simpler substructures with the implicit assumption that a comparison of the substructures is more easily and efficiently achieved. The similarity measure on the complete objects is then derived by aggregating the similarities of the constituents in a reasonable way.

This is the rationale behind the R-convolution kernels, representing a generic way to define kernels for discrete structured objects. Generally, an R-convolution kernel $k : \mathscr{G} \times \mathscr{G} \to \mathbb{R}$ can be expressed in the following form:

$$\kappa(G, G') = \sum_{g \in R^{-1}(G)} \sum_{g' \in R^{-1}(G')} \kappa(g, g') \ , \tag{4.11}$$

where $R^{-1}(G)$ denotes a decomposition of G into substructures, and κ is a kernel defined on them. With $\mathscr{G} = \mathbb{G}$, this framework can be applied to graphs. In the following, two specific instances of (4.11) are considered.

Several different kernels have been suggested in the R-convolution framework for the comparison of graphs (cf. Chapter 2), the most prominent ones being the random walk kernel (Gärtner, 2003) and the shortest path kernel (Borgwardt et al., 2005), both of which could be used for the comparison of protein binding sites. However, to utilize the full information encoded in the graph models used in this thesis, an adaptation of these kernels is necessary. In the following, these kernels are briefly introduced and expanded upon.

4.2.1.1 Random walk kernel

The random walk kernel is an R-convolution kernel which decomposes the graphs to be compared into substructures that represent walks. Random walk kernels were first introduced by Gärtner (2003) on unweighted graphs. The main idea is to decompose a graph into randomly generated walks and then count the number of identical random walks that can be found in two graphs. To calculate the random walk kernel for a given pair of graphs, it is not necessary to actually sample the walks randomly. Instead, Borgwardt et al. (2005) suggested to exploit again the concept of a product graph, albeit only on discretely labeled graphs. More precisely, one can exploit a property of the adjacency matrix $A_\times$ of the product graph $G_\times$.

For a given graph G, the number of walks of length n from node i to node j is given by $[A^n]_{i,j}$ where A^n denotes the n-th power of the adjacency matrix A. By exploiting the product graph as defined in Def. 12, $[A^n_\times]_{i,j}$ gives the number of equal walks of length n from node i to node j that occur in G_1 as well as in G_2.

Note that node labels as well as real-valued edge weights of the corresponding walks in G_1 and G_2 automatically match, as this requirement is implicitly encoded in the product graph definition by requiring $l_1(v_1) = l_2(v_2)$ for product nodes (v_1, v_2) and $|w_1(v_1, v_1') - w_2((v_2, v_2'))| \leq \varepsilon$ for product edges $((v_1, v_1'), (v_2, v_2'))$.

Thus, the random walk kernel is defined as follows:

Definition 17 *(Random walk kernel)*
Given two graphs G_1, G_2, with the product graph $G_\times = (V_\times, E_\times)$ and $A_\times$ denoting the adjacency matrix of $G_\times$. With a sequence of weights $\lambda = \lambda_0, \lambda_1, \ldots$ ($\lambda_i \in \mathbb{R}, \lambda_i \geq 0$ for all $i \in \mathbb{N}$) the random walk kernel κ_{RW} is defined by

$$\kappa_{RW}(G_1, G_2) = \sum_{i,j=1}^{|V_\times|} \left[\sum_{k=0}^{\infty} \lambda_k \cdot A_\times^k \right], \tag{4.12}$$

if such a limit exists.

Based on this definition, the random walk kernel $k_{RW}(G, G')$ can be calculated via simple matrix operations. The calculation of κ_{RW} can be done in different ways, depending on the choice of λ. Two particular choices that present themselves naturally are the geometric series and the exponential series.

Setting $\lambda_k = \lambda^k$ leads to the geometric random walk kernel:

$$\kappa_{RW}^{geo}(G_1, G_2) = \sum_{i,j=1}^{|V_{times}|} \left[\sum_{k=0}^{\infty} \lambda^k \cdot A_\times^k \right] = \sum_{i,j=1}^{|V_\times|} \left[(I - \lambda A_\times)^{-1} \right]_{ij} . \tag{4.13}$$

For the purposes of this thesis, λ is set to $\lambda_k = \lambda^k = \frac{1}{a}^k$ with $a \geq \max_{v \in V}\{deg(V_\times)\}$. The calculation of κ_{RW}^{geo} requires the inversion of $(I - \lambda A_\times)$, which is an $M^2 \times M^2$ matrix with $M = \max\{|V_1|, |V_2|\}$. Since matrix inversion has a cubic effort in the size of the matrix, the complexity amounts to $\mathcal{O}(M^6)$.

Alternatively, the exponential random walk is obtain by setting $\lambda_k = \frac{\beta^k}{k!}$, thus exploiting the exponential series:

$$\kappa_{RW}^{exp}(G_1, G_2) = \sum_{i,j=1}^{|V_{times}|} \left[\sum_{k=0}^{\infty} \frac{(\beta \cdot A_\times)^k}{k!} \right] = \sum_{i,j=1}^{|V_\times|} \left[e^{\beta \cdot A_\times} \right]_{i,j} . \tag{4.14}$$

Again, the complexity amounts to $\mathcal{O}(M^6)$, since the calculation of κ_{RW}^{exp} involves the diagonalization of the matrix $A_\times$, which is of cubic complexity.

4.2.1.2 Shortest path kernel

The random walk kernel implicitly considers all possible walks by definition. This might be somewhat problematic, as it introduces a certain redundancy in the similarity measure.

Moreover, the random walk kernel suffers from two problems known as *tottering* and *halting*. Tottering occurs if nodes or edges are visited repeatedly, thus attributing more weight to these nodes, respectively edges, which might lead to an overestimation of the similarity. This is especially severe for the graph models used here, as the protein binding sites are modeled by undirected graphs. As a result, a walk can even totter between the same two nodes repeatedly.

Halting refers to the phenomenon that the similarity measure is dominated by shorter walks. Walk kernels suffer from this problem due to the decay factor λ which down-weights larger walks. Thus, to ensure the convergence of the series, one inevitably inherits a bias towards shorter walks.

Another problem is the complexity of the random walk kernel. While a complexity of $\mathcal{O}(M^6)$ is of course preferable to solving an NP-complete problem, it is still rather high. Thus, as an alternative, Borgwardt and Kriegel (2005) introduced the shortest path kernel which considers only the shortest paths between any two nodes in order to reduce the number of considered graph components.

Again, for the purpose of this thesis, the following extension of the shortest path kernel is proposed, in order to make it applicable for the given graph models: Let $(v_{\phi_1}, ..., v_{\phi_k})$ denote the shortest path between two nodes $v_i, v_j \in G$ with $v_i = v_{\phi_1}$ and $v_j = v_{\phi_k}$. Let the length of the shortest path be defined by $lp(v_i, v_j)$ with:

$$lp(v_i, v_j) = \sum_{l=1}^{k-1} w(v_{\phi_l}, v_{\phi_l+1}) \quad . \tag{4.15}$$

Testing for equality on real-valued edge weights would obviously not be reasonable, due to the measurement accuracies and uncertainties. Therefore, edge lengths are discretized into bins of size 1. Now, the shortest path can be represented as a triple $sp(v_i, v_j)$ with $sp(v_i, v_j) = (l(v_i), l(v_j), lp(v_i, v_j))$. Obviously, setting a maximum edge weight is even necessary in this case, as otherwise the shortest path between two distinct nodes would always have a length of one.

In other words, the shortest path is defined by the label of the starting node, the end node and the sum of the discretized edge weights. On the one hand, discretization introduces some error tolerance in order to deal with the inherent noise associated with the edge weights. On the other hand, this also introduces another source of error.

Based on that simple representation, one can use the Dirac kernel to compare two shortest paths $sp(v_i^1, v_j^1)$ and $sp(v_i^2, v_j^2)$, with $v_i^1, v_j^1 \in V_1$ and $v_i^2, v_j^2 \in V_2$:

$$\kappa_{path}((v_i^1, v_j^1),\ p(v_i^2, v_j^2)) = \begin{cases} 1 & \text{if } sp(v_i^1, v_j^1) = sp(v_i^2, v_j^2) \\ 0 & \text{else} \end{cases} .$$

With this, the generalized shortest paths kernel is defined as follows:

$$\kappa_{SP}(G_1, G_2) = \frac{1}{C} \sum_{v_i^1, v_j^1 \in V_1} \sum_{v_i^2, v_j^2 \in V_2} \kappa_{path}(sp(v_i^1, v_j^1), sp(v_i^2, v_j^2)) \ , \tag{4.16}$$

where $C = \frac{1}{4}(|V_1|^2 - |V_1|) \cdot (|V_2|^2 - |V_2|)$ is a normalizing factor that guarantees $0 \leq \kappa_{SP}(G_1, G_2) \leq 1$ and more importantly ensures that κ_{SP} is size invariant.

To calculate the shortest path of a graph, several algorithms exist, one of the most prominent being the Floyd-Warshall algorithm (Floyd, 1962), which will be used in this case. The Floyd-Warshall algorithm has a cubic complexity. The shortest path kernels considers all shortest paths in two graphs in a pairwise fashion and compares them using (4.16), which has a complexity of $\mathcal{O}(1)$. Therefore, the calculation of κ_{SP} amounts to $\mathcal{O}(M^4)$ assuming $|V| = |V'| = M$, since M^4 comparisons have to be made. The described kernel avoids the above mentioned tottering problem. Moreover, the runtime complexity of the shortest path kernel amounts to $\mathcal{O}(M^4)$ which is more efficient than the random walk kernel.

Realizing the graph comparison approach as a local method inevitably leads to loss of information, since one neglects the overall structure of the graph. This is of course also true for the kernel methods presented above. However, in the case of the shortest path kernel, the loss of information incurred by reducing the information of the shortest path to the labels of start and end nodes and the associated path length might be too drastic. To put it differently, can the performance of the shortest path kernel be improved by utilizing the information given by the intermediate nodes and edges as well? Thus, a natural alternative would be to represent the shortest path simply as the sequence of node and edge labels that constitutes the path, i.e.,

$$sp_{full}(v_1, v_k) = (l(v_1), \lfloor w(v_1, v_2) \rfloor, \ldots, \lfloor w(v_{k-1}, v_k) \rfloor, l(v_k)) \ . \tag{4.17}$$

Obviously, using the Dirac kernel to compare two such shortest path sequences would not be reasonable, as this would result in a relatively crude "all or nothing" evaluation. Instead, since sp_{full} is a sequence of node labels and edge weights, one could instead utilize sequence analysis methods to obtain a more fine-grained measure. One possibility would be to use the Needleman-Wunsch algorithm (Needleman and Wunsch, 1970), which is usually employed for the calculation of pairwise sequence alignments, to compare two path sequences. This algorithm utilizes a scoring function based on the Levenshtein distance (Levenshtein, 1966) which can be used as a score that indicates how well two path sequences are in accordance. If a suitable scoring parameterization is used (1 for a match, 0 for a gap or mismatch), this scoring function fulfills the properties of a metric.

However, to employ the Needleman-Wunsch algorithm, an adaptation is necessary to avoid that edge weights are matched to node labels and vice versa. This can easily be realized by setting the score for a node-to-edge mapping to $-\infty$. As an additional modification, the score for each comparison of two path sequences is normalized by dividing it by the length of the largest path sequence in the graphs, which leads to an up-weighting of longer path sequences. The rationale behind this is that longer path sequences would carry more information, simply since more nodes and edges are visited and thus more of the overall topology is covered. Thereby a similarity measure in the interval $[0,1]$ is obtained which can be used in (4.16) instead of the Dirac kernel.

The downside of this variant is again an increased runtime. The total complexity amounts to $\mathcal{O}(M^3) + \mathcal{O}(M^2 \cdot M^2 \cdot M^2) = \mathcal{O}(M^6)$ since first the Floyd-Warshall algorithm is used to obtain all shortest paths and the comparison via the adapted Needleman-Wunsch algorithm (with $\mathcal{O}(n^2)$) has to be performed for $M^2 \cdot M^2$ sequences (assuming that the number of nodes in both graphs is M).

The shortest path kernel essentially avoids the problem of tottering and, at least in its simpler form, offers a better runtime behavior than the random walk kernel. However, by focusing explicitly on shortest paths, the problem of halting is still an issue, perhaps even more so, as larger paths are not only down-weighted but completely neglected. Thus, it is impossible to judge in advance, which variant will be best suited for the comparison of protein binding sites.

Both kernels in their original form have already successfully been applied on the comparison of whole proteins, although in a different problem setting. More precisely, both kernels have been used on an SSE-based graph representation of protein folds (Borgwardt et al., 2005). Whether the above kernels with the introduced extensions will be equally useful for the comparison of protein binding sites will be investigated in Chapter 5.

4.2.2 Fingerprints

The above-described methods basically represent extensions of existing methods. As outlined above, R-convolution kernels potentially offer some benefits, such as an increased runtime efficiency and presumably a greater tolerance towards structural variation compared to global approaches. They also suffer from different drawbacks as well, such as tottering and halting. Moreover, whether the R-convolution framework itself is suitable for the specific problem of protein binding site comparison remains debatable.

Instead, since a loss of information is deliberately accepted by using decomposition techniques, one could also reduce the graph representation to a feature representation of a fixed length and subsequently compare the feature vectors. For the purpose of retrieving similar graphs (respectively binding sites) from a database, it is not even necessary to realize the comparison of the feature vectors as a kernel function. Instead, a suitable distance metric might suffice.

One possibility to derive such a mapping to feature vectors is to define a set of *patterns* whose presence in the graph is checked. A pattern naturally corresponds to a subgraph in case of graph objects[3]. Each entry in the feature vector corresponds to a specific pattern and indicates whether the corresponding pattern is present in the graph. Therefore, the feature vector serves as a *fingerprint* of the original graph.

Based on such a fingerprint representation, a plethora of feature-based comparison methods become available. If the fingerprint comparison is realized by a kernel function, one can obtain a very simple kernel outside the R-convolution framework. In the following, such a fingerprint approach is introduced for the comparison of protein binding sites.

4.2.2.1 Crisp fingerprints

Obviously, the fingerprint representation depends on the predefined set of patterns. This raises the question, which types of subgraphs should be used, as in principle any subgraph could be chosen, including walks and paths on which the previous methods have focused. On the one hand, one would need patterns that are not too complex, since this would increase the possibility that a certain pattern would only be present in a handful of graphs, if it occurs at all. On the other hand, the patterns need to be specific enough to be of discriminative value. From a technical point of view, it would be advisable to keep the

[3]Alternatively, one could also include other features, e.g., based on general attributes such as graph size, node density and so on, although these would be global instead of local features.

patterns simple to limit fingerprint size and keep the calculations feasible. Since the construction of the fingerprints involves checking for the presence of each pattern, this means that some kind of isomorphism test needs to be performed for each entry. Hence, the smaller the patterns, the more efficient the runtime performance.

Given these considerations, subgraphs of size three were chosen as patterns to ensure a high runtime efficiency, which is one major motivation to use local methods in the first place. Obviously, not all possible subgraphs of size three can be considered, given that the graphs used in this thesis feature real-valued edge weights and thus an infinite number of possibilities exist. Also, sampling patterns from existing graphs would not be reasonable, as the result would strongly depend on the graphs chosen for the sampling. Instead, one can again resort to discretization by considering n distinct node labels and k distinct edge weights, which gives rise to a finite number of possible patterns given by:

$$N(n,k) = \binom{n}{3} \cdot k^3 + n(n-1) \cdot k \cdot \binom{k+1}{2} + n \cdot \binom{k+2}{3} \tag{4.18}$$

This is easily verified since only three cases can occur:

1. The pattern contains three different node labels: There are $\binom{n}{3}$ possibilities to choose three distinct labels. As this also uniquely identifies the edges, there are k^3 possibilities for the edge labels.

2. The pattern contains two equal node labels that differ from the third: There are $n(n-1)$ possibilities to choose two distinct labels, one for the identically labeled nodes and one for the third. In this case a graph with the edges emanating from the uniquely labeled node swapped would be isomorphic. To account for this, one can sort these edges according to their weight which would map isomorphic patterns to the same representation, which leads to $k \cdot \binom{k+1}{2}$ possible edge combinations.

3. The pattern contains only identical node labels: There are n possibilities to choose the node label. Again, to find a unique representation that accounts for isomorphism, all edges can be sorted which leads to $\binom{k+2}{3}$ possible edge combinations.

To test for the presence of a given pattern, an ε-threshold can again be employed in analogy to the GAVEO approach introduced in Section 4.1.1. A pattern t_i is contained in a graph G, if there is a subgraph in G_s which is ε-isomorphic to t_i.

Alternatively, one could use a simple binning strategy, by partitioning the set of real-valued edge weights into several intervals of the bin size b. In this case, instead of designating a certain set of discrete edge labels and a tolerance threshold ε, a bin size b is used to specify the fingerprints. Accordingly, a pattern t_i is contained in a graph G, if there is a subgraph in G_s, whose edge weights fall into the bins specified by the edge labels t_i.

In both cases, a fingerprint is defined in the following way:

Given a graph G, let

$$f_G = \left(G \sqsupseteq t_1, G \sqsupseteq t_2, \ldots, G \sqsupseteq t_{N(n,k)}\right) \in \mathbb{N}^{N(n,k)}$$

where $\{t_1, \ldots, t_{N(n,k)}\}$ is the set of all non-isomorphic patterns of size three defined by fixed sets of node labels and edge weights, numbered in an arbitrary but fixed order. The predicate $G \sqsupseteq t_i$ tests whether t_i is contained in G and returns the number of occurrences. Again, setting an upper limit δ for edge weights is necessary here and has the effect of limiting the size of the fingerprints, which positively affects runtime efficiency.

To improve runtime performance during the construction of such fingerprint vectors, one can make use of a hashing function based on canonical forms of the given patterns, instead of employing a brute-force approach. In this work, the canonical forms are based on the above distinctions between the types of possible patterns of size three[4]:

1. All node labels are identical. In this case, the canonical form is given by the node label followed by the edge lengths in increasing order.

2. Two nodes have an identical label. The canonical form starts with the node label that appears once in the graph followed by the label that appears twice, the edge weight between the nodes with the same label, and finally the remaining two edge weights in increasing order.

3. All nodes have different labels. The canonical form is then defined by the three occurring labels, sorted in a lexicographic order, the edge length between the first and the second, the second and the third, and finally the first and the third node.

All three cases are illustrated by an example in Fig. 4.6. We denote the set of canonical forms by Γ. The above representation enables the definition of a bijective function $i : \Gamma \rightarrow \{1, \ldots, N(n,k)\} \subset \mathbb{N}$ assigning a unique number to each form and, therefore, subgraph of size 3.

Using this mapping, the calculation of the fingerprint vector for a graph $G = (V, E)$ can be done in a more efficient way by enumerating all subgraphs of size 3 in G. For each subgraph g_i of size 3 in G, the transformation to its canonical form σ_i is performed (in time

[4]Though in principle also other conventions are possible.

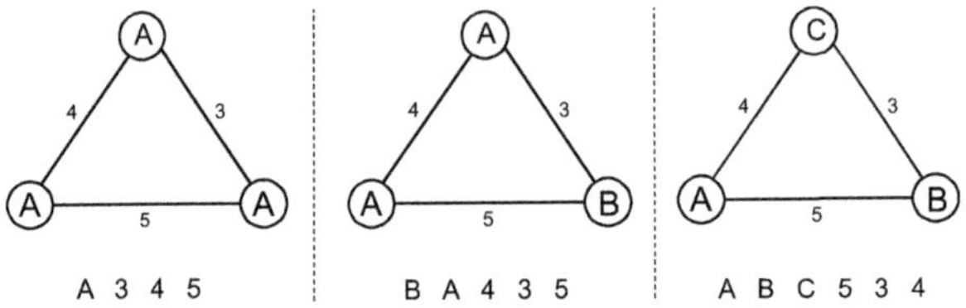

Figure 4.6: The three possible cases that can occur: all labels identical, two labels identical and all labels unique.

$\mathcal{O}(1)$) and the function $i(\sigma_i)$ is evaluated to determine the position of g_i in the fingerprint vector (in time $\mathcal{O}(1)$). Finally the entry at this position in the vector is incremented by one. Doing this for all $\binom{M}{3} = \mathcal{O}(M^3)$ subgraphs of size 3 leads to a runtime complexity of $\mathcal{O}(M^3)$.

Given such a feature representation, the comparison of two graphs G_1 and G_2 is transferred to the comparison of their respective fingerprint vectors f_{G_1} and f_{G_2}. For this purpose, different distance measures can be employed, one of the simplest being the Hamming distance.

Hamming fingerprints

If one is merely interested in the presence or absence of a pattern, a simple distance function can be devised based on the Hamming distance. For each pattern, the simultaneous absence or presence in both graphs is rewarded and aggregated to the following similarity measure:

$$k_{FPH}(G_1, G_2) = \frac{1}{N(n,k)} \sum_{i=1}^{N(n,k)} k_\delta([f_{G_1}]^i, [f_{G_2}]^i) \ , \tag{4.19}$$

where $[f_{G_1}]^i$ denotes the i-th entry in the vector f_{G_1}, and

$$k_\delta(x,y) = \begin{cases} 1 & (x > 0 \wedge y > 0) \vee (x = 0 \wedge y = 0) \\ 0 & \text{otherwise} \end{cases} \ . \tag{4.20}$$

Jaccard Fingerprints

A potential disadvantage of using the Hamming distance is the fact that it does not only reward the simultaneous presence of a pattern, but also its absence. This is somewhat counterintuitive, since the absence of a certain pattern can obviously not hint at a shared functionality of the corresponding binding pockets. Therefore, an alternative measure from the field of set theory can be employed that avoids this problem. By utilizing the well-known Jaccard coefficient

$$J(A,B) = \frac{A \cap B}{A \cup B} , \tag{4.21}$$

an alternative similarity measure can be obtained:

$$k_{FPJ}(G,G') = \frac{\sum_{i=1}^{N(n,k)} \min([f_{G_1}]^i, [f_{G_2}]^i)}{\sum_{i=1}^{N(n,k)} \max([f_{G_1}]^i, [f_{G_2}]^i)} . \tag{4.22}$$

Of course, a plethora of other possible distance measures could also be used instead, for example cosine similarity, the Minkowski metric, etc. However, for the sake of brevity, the focus will be on the introduced methods as a proof of concept.

4.2.2.2 Fuzzy fingerprints

The previously introduced approach relies on the discretization of edge weights to define the distinct patterns. A real-valued edge weight corresponds to the discrete edge weight of a given pattern, if the difference between these edge weights is within an ε range or a certain interval. This effectively enforces that each weight actually encountered in a graph is assigned exclusively to one certain edge label.

Conceptually, this is a potential disadvantage of the proposed method due to the problem of discontinuity, which might cause a number of problems. For example, cases can occur where similar subgraphs are considered distinct, (i.e., they are considered to correspond to different patterns) due to a minor difference in edge weight, whereas more dissimilar subgraphs are attributed to the same pattern and thus are considered equal. An illustrating example is given in Fig. 4.7. Moreover, a slight deviation of an edge weight can lead to the absence of a feature if the ε threshold is exceeded.

One possible way to avoid these issues is to resort to *fuzzy* discretization. In fuzzy set theory, a fuzzy partition of a domain X (also called the *universe of discourse*) is defined by a finite family of fuzzy subsets $F_{i_1}, F_{i_2}, \ldots, F_{i_k}$ of X such that $\sum_{i=1}^{k} F_{i_k}(x) > 0$ for all $x \in X$. Additionally, one often requires that $\sum_{i=1}^{k} F_i(x) = 1$ for all $x \in X$, though this is not a necessity.

Definition 18 *(Fuzzy set)*

A fuzzy set F is defined by a function μ on the reference domain X to the unit interval $[0,1]$: $F := \{(x, \mu(x)) | x \in X\}$. The function μ is called a membership function.

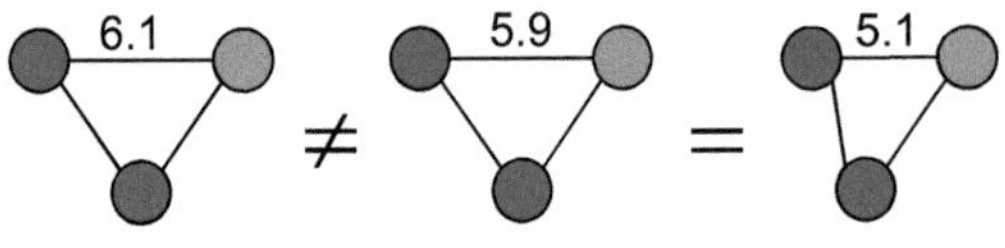

Figure 4.7: Example of a discontinuity problem. Given that edge weights are separated into the intervals $[5,6[$ and $[6,7[$, the left and the center graph would be considered dissimilar, while the center and right graph would correspond to the same pattern. This is clearly counterintuitive, since the left and center graph show a much lower difference in edge lengths.

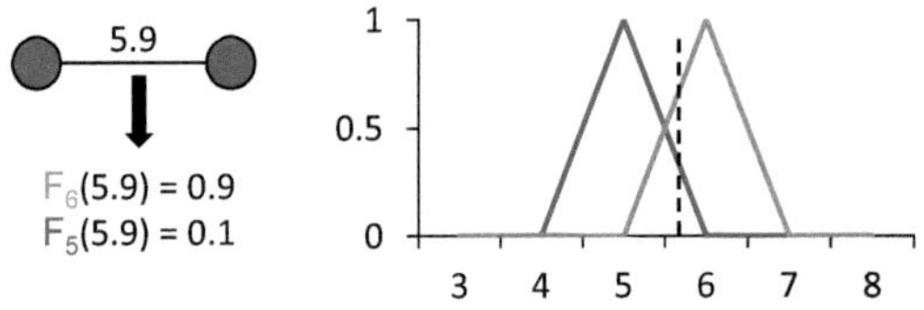

Figure 4.8: Two fuzzy sets F_5 and F_6 defined by their membership functions. The real-valued edge weight 5.9 corresponds to the fuzzy set F_5 with a degree of 0.1 and to the fuzzy set F_6 with a degree of 0.1.

For each $x \in X$, $\mu(x)$ returns the *degree of membership* to the fuzzy set F, i.e., every $x \in X$ belongs to the fuzzy set F to a certain degree specified by μ.

In this work, a fuzzy partition of the domain $X = \mathbb{R}_0^+$ is realized by fuzzy sets F_i, with

$$F_i(x) = \max\{0, 1 - |\frac{x-i}{\eta}|\} \quad , \tag{4.23}$$

thus realizing a triangular membership function, with η denoting a radius around the label i that specifies the support of the fuzzy set. In fuzzy set theory, the support represents the set of all points in the universe of discourse with a membership degree $\mu > 0$. In this case, the support is given by $S_i = \{x | x \in]i-\eta, i+\eta[\}$.

Intuitively, the fuzzy set F_i can be interpreted as the fuzzy subsets of numbers that are "approximately equal to i". Thus, a real-valued edge weight belongs to several fuzzy sets up to a certain degree, as illustrated in Fig. 4.8.

To realize a fuzzy fingerprint, the concept of a pattern is altered by replacing the discrete edge weights by fuzzy sets F_i. As a result, a subgraph g of a graph G with real-valued edge weights can correspond to multiple patterns t up to a certain degree of membership which can be interpreted as a "degree of isomorphism".

This degree of membership is defined in the following way: Let a_i denote the label of the i-th node in g, and x_{ij} the weight of the edge connecting the nodes i and j. Let b_i be the label of the i-th node in t, and F_{ij} the fuzzy set representing the weight of the edge between node i and node j. The degree of isomorphism of t and g, denoted $[t \sim g]$, is then given by

$$[t \sim g] = \max_{\pi \in S_3} \begin{cases} \min\{F_{12}(y_{12}), F_{13}(y_{13}), F_{23}(y_{23})\} & \text{if } M(\pi) \\ 0 & \text{otherwise} \end{cases} \tag{4.24}$$

where $y_{ij} = x_{\pi(i),\pi(j)}$, S_3 is the set of all permutations $\pi : \{1,2,3\} \to \{1,2,3\}$. $M(\pi)$ is true if

$$(a_1 = b_{\pi(1)}) \wedge (a_2 = b_{\pi(2)}) \wedge (a_3 = b_{\pi(3)})$$

and false otherwise.

The degree to which a pattern t is present in the graph G is derived by taking the maximum over all subgraphs g of size three:

$$[G \sqsupseteq t] = \max_{g} \, [t \sim g] \ . \tag{4.25}$$

This value defines the entry for the pattern t in the fuzzy fingerprint vector f_G derived from G. The resulting feature vector contains entries from the unit interval $[0,1]$ instead of a binary value that indicates presence or absence of a pattern. Intuitively, this is a more intuitive model than a relatively crude all-or-nothing approach that potentially enhances the discriminative power of the fingerprint approach. Moreover, it circumvents the problem of discrete boundaries.

From a technical point of view, fingerprint construction occurs in analogy to the non-fuzzy approach by utilizing a hash map as described in the previous section. The notable difference is that, instead of incrementing the entry for an encountered patten, the entry is simply updated according to (4.25).

Having derived an alternative fingerprint representation, the comparison of two fingerprints f_{G_1} and f_{G_2} is realized by defining a similarity measure based on the Jaccard coefficient (4.21), with the difference that the logical operators from set theory are generalized by their counterparts in fuzzy set theory. Thus, union and intersection are replaced by fuzzy t-norms and t-conorms, respectively, resulting in the following similarity measure:

$$k_{FFP}(G_1, G_2) = \frac{\sum_{i=1}^{N(n,k)} \top([f_{G_1}]^i, [f_{G_2}]^i)}{\sum_{i=1}^{N(n,k)} \bot([f_{G_1}]^i, [f_{G_2}]^i)} .$$

In the experiments in Chapter 5, the t-norm will be realized as a maximum operator and the t-conorm as a minimum operator: $\top(a,b) = \min(a,b)$ and $\bot(a,b) = \max(a,b)$.

Generally speaking, it is not clear in advance, which of the described variants will be more useful in the context of protein binding site comparison, though conceptually the fuzzy fingerprint approach appears to be more intuitive. This question will be addressed in the next chapter.

4.3 Semi-global graph comparison

In the previous section, several local graph comparison methods have been introduced as alternatives to global graph comparison. These approaches derive a similarity measure of graphs by comparing local graph features, e.g., subgraphs of a specific type. As outlined above, approaching the graph comparison on a local scale offers some potential benefits, such as an improved runtime efficiency. Moreover, from a conceptual point of view, local methods based on decomposition techniques will be much less affected by conformational differences and inaccuracies in the measurements, simply as the overall structure becomes less important and the comparison process is less strict. As argued above, this can be beneficial, especially if no overall similar graph topology exists and only subpockets of a binding site show some structural resemblance.

On the other hand, using such approaches inevitably bears the risk of producing a high similarity for graphs whose overall topology is different, due to the loss of information caused by decomposing the graph into substructures. In fact, decompositions of this type are typically not bijective, i.e., the complete graph cannot be recovered from the components. A simple illustration is shown in Fig. 4.9. The two graphs shown there are quite different in terms of their overall topology. Yet, they are decomposed into the same set of components (subgraphs of size two). Thus, a local method operating on these components will produce a high degree of similarity.

The question remains, whether this loss of information is critical, i.e., whether the derived similarity measures are still capable of discriminating different functional protein classes. Especially in cases where a global structural similarity exists, neglecting the overall topology of the graphs might limit the usefulness of these approaches.

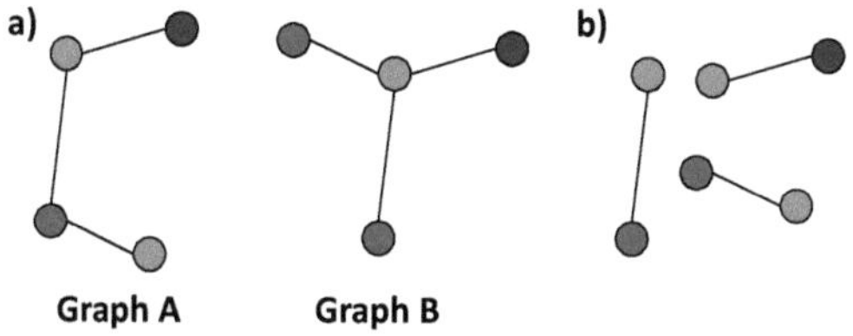

Figure 4.9: Two graphs that are quite different in terms of graph topology (a). Yet, their decomposition into subgraphs of size two yields the same set of components (b).

Another disadvantage is the lack of interpretability compared to the approaches presented in Section 4.1. While this is not problematic for retrieving possibly similar structures from a database, it does become important when studying the functional mechanisms of a certain protein class. In this case, one is typically interested in identifying the regions that are present in all or most functionally related structures, as these regions most likely play a major role for protein function (e.g., catalytic triads).

Thus, both purely local and purely global graph comparison methods theoretically have some merits as well as some limitations for the application at hand. In advance it is hard to tell in which case the benefits outweigh the downsides of the respective methods, a question that will be further addressed in Chapter 5.

However, another question comes to mind: having recognized the merits and limitations of purely global or purely local graph comparison, is it possible to combine the advantages of both the global and the local principle while at the same time avoiding their disadvantages? A possible solution to this conundrum is again inspired by the realm of sequence alignment. In addition to global and local sequence alignments, semi-global sequence alignments have also been widely used for different tasks. Thus, a natural idea is to calculate a graph alignment as introduced in Chapter 3 in a semi-global manner. In the following, a semi-global approach to the comparison of graphs called SEGA (SEmi-global Graph Alignment) will be introduced.

4.3.1 SEGA - SEmi-global Graph Alignment

SEGA represents a semi-global strategy for the calculation of graph alignments as defined in Chapter 3 in the sense that it shares properties with both local and global methods. SEGA again establishes a correspondence between nodes, like the global approaches introduced earlier. However, instead of optimizing the alignment globally, SEGA assembles it from local comparisons. Since such local comparisons bear the risk of producing ambiguities, for example when local subgraphs occur several times in the graph, SEGA resorts to the overall graph topology to resolve these ambiguities.

The rationale behind this idea is that, by resorting to local comparisons, the approach becomes more tolerant towards structural variation, while at the same time avoiding the risk of making arbitrary assignments of nodes by resorting to global information if necessary.

Since the goal is to establish a correspondence between nodes of different graphs, a measure of similarity between these nodes is needed to determine which nodes are most similar to each other and thus should be mutually assigned. This is realized by comparing nodes based on their labels and their immediate surroundings, the *node neighborhood*. This neighborhood is defined by the closest neighboring nodes and the edges connecting them. The local comparison is realized by comparing fully connected subgraphs of a given size, with the actual nodes of interest as center nodes. For binding pockets, this corresponds to the comparison of pseudocenters and the spatial constellation of physicochemical properties in close proximity of these centers.

Contrary to other approaches, SEGA does not aim for the identification of completely matching substructures. Instead, the goal is to obtain an estimation of the geometric similarity of two node neighborhoods which is realized by comparing triplets of pseudocenters that constitute such a neighborhood. More precisely, SEGA considers all triplets (or triangles) containing the center node and two of the neighborhood nodes to capture the spatial positioning of the center node relative to its neighbors. By deriving a mutual assignment of similar triangles and summing up the number of matches, an intuitive measure of the node similarity is derived that can be used to construct a local distance matrix.

In a second step, this matrix is used to construct a pairwise graph alignment. Essentially, this can be formulated as a weighted optimal assignment problem, which can be solved by a number of different approaches, most notably the Hungarian algorithm (Kuhn, 2005). However, in a typical optimal assignment setting, the goal would be to optimize a cost function, usually the sum of costs associated with the individual assignments. Here, this cost is given by the entries of the local distance matrix, i.e., the cost of the local neighborhood similarity.

However, for the application at hand, minimizing the sum of all costs is possibly not the best choice. In doing so, cases can occur, where highly similar nodes are not mutually assigned in order to minimize the overall cost. Arguably, this is not desirable, as the occurrence of a pair of nodes with (nearly) identical neighborhoods in two different binding sites will rarely be encountered by chance, depending on the neighborhood size. If such a nearly identical neighborhood is encountered, it is most likely due to a similar enzymatic mechanism and therefore more meaningful than pairs with higher distances. Consequently, such occurrences should be considered first.

The SEGA algorithm thus works as follows: In a first step, the local distance matrix is calculated, based on the neighborhood similarity. In a second step, the distance matrix is used to derive a mutual assignment of nodes in an incremental way, starting with the most similar nodes. Conceptually, this can be regarded as a divide-and-conquer strategy, since the problem of finding the correspondence of whole graphs is reduced to solving multiple correspondences of subgraphs, i.e., the neighborhoods, and then combining the solutions to a global one. In the following, the algorithm is described formally.

4.3.1.1 Neighborhood distance measure

Assume that two input graphs $G_1 = (V_1, E_1)$ and $G_2 = (V_2, E_2)$ with $|V_1| = n$ and $|V_2| = m$ are given. Contrary to the other methods introduced above, SEGA uses complete graphs as input, which ascertains that for each node the closest neighboring nodes can always be retrieved. In a first step, the local distance matrix

$$D = (d_{ij})_{1 \le i \le n, 1 \le j \le m} \tag{4.26}$$

of dimensionality $n \times m$ is constructed. The entry d_{ij} corresponds to the distance between the nodes $v_i^{(1)} \in V_1$ and $v_j^{(2)} \in V_2$ ($1 \le i \le n$, $1 \le j \le m$), which can be interpreted as a degree of dissimilarity that is inversely related to a corresponding similarity degree

$$s_{ij} = sim(v_i^{(1)}, v_j^{(2)}) \ . \tag{4.27}$$

The similarity (4.27) between two nodes $v_i^{(1)}$ and $v_j^{(2)}$ is defined in terms of the similarity of their respective neighborhoods. For a given (center) node $v_c \in V$, let $N(v_c, n_{neigh}) \subseteq V$ consist of the closest n_{neigh} nodes in V, i.e., those nodes having the smallest Euclidean distance from v_c. The neighborhood of v_c is then defined by the set $\mathbf{N}(v_c, n_{neigh})$ of all triangles $\{u, v_c, w\}$ with $u, w \in N(v_c, n_{neigh})$, $u \neq w$ (see Fig. 4.10).

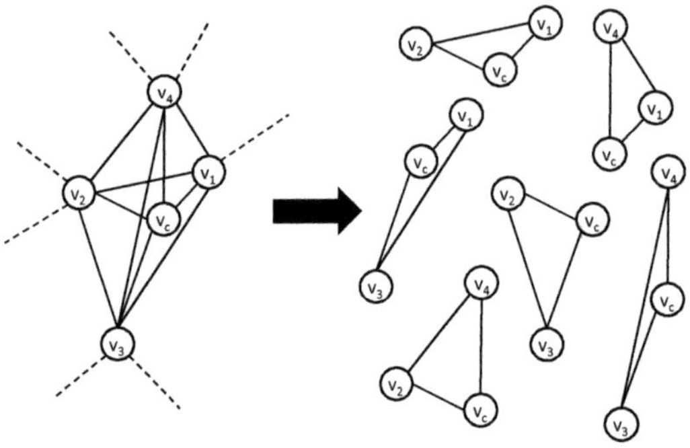

Figure 4.10: Decomposition of the neighborhood of node v_c with $n_{neigh} = 4$. The subgraph defined by the n_{neigh} nearest nodes is decomposed into triangles containing the center node v_c.

Let

$$t^{(1)} = \{v_1^{(1)}, v_2^{(1)}, v_3^{(1)}\} \in \mathbf{N}(v_i^{(1)}, n_{neigh}),\ v_1^{(1)} = v_i^{(1)},$$
$$t^{(2)} = \{v_1^{(2)}, v_2^{(2)}, v_3^{(2)}\} \in \mathbf{N}(v_j^{(2)}, n_{neigh}),\ v_1^{(2)} = v_j^{(2)},$$

be two triangles from the neighborhoods of nodes $v_i^{(1)} \in V_1$ and $v_j^{(2)} \in V_2$, respectively. Two such triangles are considered a *match* if a mapping $\phi : t^{(1)} \to t^{(2)}$ exists with either

$$\phi(v_1^{(1)}) = v_1^{(2)},\quad \phi(v_2^{(1)}) = v_2^{(2)},\quad \phi(v_3^{(1)}) = v_3^{(2)},$$

or

$$\phi(v_1^{(1)}) = v_1^{(2)},\quad \phi(v_2^{(1)}) = v_3^{(2)},\quad \phi(v_3^{(1)}) = v_2^{(2)},$$

and for which the following conditions hold:

$$(i) \quad \ell(v_2^{(1)}) = \ell(\phi(v_2^{(1)})), \quad \ell(v_3^{(1)}) = \ell(\phi(v_3^{(1)})),$$

$$(ii) \quad \max \left\{ \begin{array}{l} |e(v_1^{(1)}, v_2^{(1)}) - e(\phi(v_1^{(1)}), \phi(v_2^{(1)}))|, \\ |e(v_2^{(1)}, v_3^{(1)}) - e(\phi(v_2^{(1)}), \phi(v_3^{(1)}))|, \\ |e(v_3^{(1)}, v_1^{(1)}) - e(\phi(v_3^{(1)}), \phi(v_1^{(1)}))| \end{array} \right\} \leq \varepsilon$$

Again, the parameter $\varepsilon \geq 0$ is used as tolerance threshold determining the maximally allowed deviation of edge lengths. Roughly speaking, two triangles match, if a superposition preserving node labels and edge weights exists. The only exception concerns the center nodes $v_i^{(1)}$ and $v_j^{(2)}$: These two nodes are necessarily assigned to each other, but their labels can be different. This is done to allow for the construction of *approximate* graph alignments that may contain mismatches (mutually assigned nodes with different label), similar to the error-tolerant GAVEO approach introduced in Section 4.1.1.

As has been stated earlier, the type of a pseudocenter (and accordingly a node label) can change due to point mutations in the protein binding pocket. By allowing for mismatches, an altered pseudocenter can still be assigned to its non-mutated counterpart, provided the corresponding neighborhoods remain roughly the same. By applying this approximate matching technique one can account for biological variation caused by mutations, as well as structural variation affecting the distance between the pseudocenters.

The similarity (4.27) between two nodes $v_i^{(1)} \in V_1$ and $v_j^{(2)} \in V_2$ is now defined by the maximal number of matching triangles from $\mathbf{N}(v_i^{(1)}, n_{neigh})$ and $\mathbf{N}(v_j^{(2)}, n_{neigh})$ that can be assigned in a mutually exclusive way. That is, each triangle from $\mathbf{N}(v_i^{(1)}, n_{neigh})$ can only be matched with at most one triangle from $\mathbf{N}(v_j^{(2)}, n_{neigh})$, and vice versa. Note that

$$0 \leq s_{ij} \leq s_{max} = \frac{n_{neigh}(n_{neigh} - 1)}{2} \quad . \tag{4.28}$$

To determine s_{ij}, an optimal assignment problem has to be solved. To this end, the well-known Hungarian algorithm (Kuhn, 2005) is applied to the $s_{max} \times s_{max}$ matrix whose entry at position (k, l) is 0 if the k-th triangle in $\mathbf{N}(v_i^{(1)}, n_{neigh})$ can be matched with the l-th triangle in $\mathbf{N}(v_j^{(2)}, n_{neigh})$; otherwise, the entry is 1. The Hungarian algorithm is a combinatorial optimization algorithm that solves an optimal assignment problem based on a cost matrix as input, with a time complexity of $O((s_{max})^3)$. This is done by computing the cost of a cost-minimal assignment and returning $s_{max} - sim(v_i^{(1)}, v_j^{(2)})$, i.e., the number of triangles that could not be matched. This value defines the distance between $v_i^{(1)}$ and

$v_j^{(2)}$, i.e.,

$$d_{ij} = \left(s_{max} - sim(v_i^{(1)}, v_j^{(2)})\right)^2 \quad . \tag{4.29}$$

As a side remark, one should note that this assignment problem need not be solved by algorithms suitable for weighted optimal assignment problems. Instead, one could solve this problem more efficiently, for example by using the Hopcroft-Karp algorithm (Hopcroft and Karp, 1973). However, due to the small values of n_{neigh}, this does not affect the runtime complexity of the complete algorithm as the calculation with either algorithm is quite efficient in practice. The Hungarian algorithm will later be needed again to solve a weighted optimal assignment problem.

As local distance measure, the squared number of non-matching triangles is used (4.29). As will be seen in the next section, taking the squared number of non-matching triangles will not affect the calculation of the graph alignment, i.e., one could also use the difference between s_{max} and $sim(v_i^{(1)}, v_j^{(2)})$ directly. However, the cost associated to a certain assignment of nodes will later be used to derive a quality measure for the alignment. Squaring the difference (4.29) will serve to increase the influence of node assignments with a nearly perfect accordance regarding the spatial constellation of the associated node neighborhoods. Once the distance matrix D is derived, a graph alignment is calculated in a second step.

4.3.1.2 Deriving a global alignment

The result of the above computation is an $n \times m$ distance matrix (4.26). This matrix is used as an input for the second step of SEGA, which seeks to find an optimal mutual assignment of nodes from V_1 and V_2, respectively. Since d_{ij} can be considered as the cost of assigning nodes $v_i^{(1)}$ and $v_j^{(2)}$ to each other, this problem can again be formulated as an optimal assignment problem, finding the assignment with the minimal sum of costs. As this represents a weighted optimal assignment problem, the Hungarian algorithm can be used to solve this problem.

However, a solution thus obtained, even if being optimal in the sense of minimizing the total cost of node assignments, will usually not provide a reasonable alignment with respect to the overall graph topology. The problem is that the Hungarian algorithm does not take the spatial relationships between the nodes into consideration. As a result, assignments of nodes with the same associated cost will be treated as equivalent, regardless whether the nodes are located in corresponding regions of the compared binding pockets or in completely different regions.

Moreover, such ambiguities will be resolved in an arbitrary way. In fact, due to the nature of the underlying distance matrix, whose entries are discrete values from a fixed set of possible costs given by squaring the integers between 0 and (s_{max}), it is likely that a cost-minimal solution is not unique. The Hungarian algorithm will simply pick one from the set of all cost-optimal solutions, which is not necessarily in agreement with the overall topology.

Additionally, even non-ambiguous highly affine nodes might not be assigned to each other. This can occur if an alternative assignment of such nodes would break up another, highly expensive assignment of two dissimilar nodes, resulting in a situation, where two mediocre node assignments are preferred to one high and one low scoring assignment. Intuitively, this is not desirable, as a highly affine pair of nodes is much more likely to represent a conserved and thus functionally important region of the binding pocket. While this problem is somewhat mitigated by using the squared number of non-matching triangles as local distance measure, creating a preference for highly similar neighborhoods, it is nevertheless possible.

A high neighborhood similarity is only achieved if a conserved common substructure exists. The size of this substructure directly affects the number of highly affine node pairs and the most similar pairs should always correspond to nodes located in the center of the associated common subgraph, where a high neighborhood similarity is most likely. Hence, one should consider such nodes first. Consequently, SEGA will assemble an assignment by starting with the nodes exhibiting the lowest observed distance value and then incrementing through the possible distance values (note that only a fixed number of values are possible), making all possible non-ambiguous assignments before advancing to the next level.

If more than one assignment for a given node is possible with the same cost, SEGA resorts to global information from graph topology to resolve such ambiguities. More specifically, an initial seed solution is constructed in the form of a partial assignment of nodes which will serve as a reference frame. To this end, only nodes $v_i^{(1)} \in V_1$ and $v_j^{(2)} \in V_2$ having a distance of 0 and, hence, being highly affine, are considered. If such nodes exist and can be mutually assigned without ambiguities, these assignments are realized. With

$$f_c(v_i^{(1)}) = \{\, v_j^{(2)} \in V_2 \,|\, d_{ij} \leq c \,\} \ ,$$

$$g_c(v_j^{(2)}) = \{\, v_i^{(1)} \in V_1 \,|\, d_{ij} \leq c \,\} \ ,$$

(where e.g., $f_c(v_i^{(1)})$ denotes the set of vertices in G_2 whose distance to $v_i^{(1)}$ is not greater than c) those pairs $v_i^{(1)}$ and $v_j^{(2)}$ satisfying $f_0(v_i^{(1)}) = \{v_j^{(2)}\}$ and $g_0(v_j^{(2)}) = \{v_i^{(1)}\}$ are assigned, as they represent unambiguous choices for constructing the seed solution. Those nodes $v_i^{(1)}$ with $|f_0(v_i^{(1)})| > 1$ (and $v_j^{(2)}$ with $|g_0(v_j^{(2)})| > 1$) are not yet assigned, as for these nodes multiple conflicting assignments are possible. Such conflicting choices are later resolved by drawing on the seed solution as reference frame.

The seed solution thus obtained must satisfy the constraint that the set of mapped points for each graph contains a basis of $\mathbb{R}^3$ to determine the relative position of a new node in three-dimensional space in an unambiguous way. To ensure this, at least four pairs of points are needed, provided these points can be used to define a spanning set of vectors for $\mathbb{R}^3$ that are linearly independent. If this condition is not met, a sufficient number of candidate pairs is collected by relaxing the distance constraint, i.e., a maximal local distance $c > 0$ is allowed.

If even the seed solution cannot be constructed unambiguously, the following strategy is employed: Let $S_1 \subseteq V_1$ and $S_2 \subseteq V_2$ denote the nodes occurring in these candidates. SEGA then constructs all possible candidate assignments

$$\left((s_1^{(1)}, s_1^{(2)}), (s_2^{(1)}, s_2^{(2)}), (s_3^{(1)}, s_3^{(2)}), (s_4^{(1)}, s_4^{(2)}) \right) \subseteq S_1 \times S_2$$

of size four that represent a unique three-dimensional geometry and are unambiguous in the sense that $s_i^{(2)} \in f_c(s_i^{(1)})$ and $s_j^{(2)} \notin f_c(s_i^{(1)})$ as well as $s_i^{(1)} \in g_c(s_i^{(2)})$ and $s_j^{(1)} \notin g_c(s_i^{(2)})$ for all $1 \leq i \neq j \leq 4$. As final seed solution, the candidate minimizing the spatial deviation

$$\sum_{1<i<j<4} \left| e(s_i^{(1)}, s_j^{(1)}) - e(s_i^{(2)}, s_j^{(2)}) \right|$$

is selected to match the candidates that are most similar in terms of geometry.

Now, suppose a current seed in the form of a partial alignment to be given. There may still be the problem that some nodes could not be assigned unambiguously. To solve this problem, one can again formulate an optimal assignment problem, this time augmented by drawing upon global information. In the k-th iteration, nodes having a distance of at most c_k are assigned, where c_k is the k-th smallest cost value in the matrix D. More specifically, let $W_1 \subset V_1$ ($W_2 \subset V_2$) denote the set of nodes from V_1 (V_2) that have already been assigned in a previous iteration. Moreover, let

$$U_1^k = \{v_i^{(1)} \in V_1 \,|\, f_{c_k}(v_i^{(1)}) \neq \emptyset\} \setminus W_1 \ ,$$
$$U_2^k = \{v_j^{(2)} \in V_2 \,|\, g_{c_k}(v_j^{(2)}) \neq \emptyset\} \setminus W_2 \ .$$

Then a (partial) assignment of nodes in U_1^k and U_2^k is derived by applying the Hungarian algorithm to a cost matrix defined as follows. The matrix contains an entry for each pair of nodes $v_i^{(1)} \in U_1^k$ and $v_j^{(2)} \in U_2^k$. If $v_j^{(2)} \notin f_{c_k}(v_i^{(1)})$, the corresponding cost value is set to a sufficiently high constant C (indicating that these two nodes should not be assigned). Otherwise, the cost value is determined by resorting to information from the (global) graph structure, by comparing the position of $v_i^{(1)}$ relative to the current seed nodes W_1 with the position of $v_j^{(2)}$ relative to W_2. More precisely, the cost is defined by

$$\sum_{q=1,2,\ldots,|W_1|} \left| |v_i^{(1)} - w_q^{(1)}| - |v_j^{(2)} - w_q^{(2)}| \right| \ ,$$

where $w_q^{(1)}$ and $w_q^{(2)}$ denote, respectively, the q-th node in W_1 and W_2 (which are mutually assigned), and $|v_i^{(1)} - w_q^{(1)}|$ is the Euclidean distance between $v_i^{(1)}$ and $w_q^{(1)}$.

Applying the Hungarian algorithm yields again a cost-minimal assignment. If $v_i^{(1)}$ and $v_j^{(2)}$ participate in this assignment, i.e., have been assigned to each other, $v_i^{(1)}$ is added to W_1 and $v_j^{(2)}$ to W_2 if $v_j^{(2)} \in f_{c_k}(v_i^{(1)})$, i.e., if the corresponding cost value is smaller than C. Intuitively, the main idea behind this step is to choose only those possible node assignments from all ambiguous choices, for which the corresponding nodes are roughly oriented in the same manner towards the reference frame given by the seed solution or at least show the least deviation.

This procedure iterates until all nodes of one graph are assigned, or until a predefined upper cost value c_{max} has been reached, with remaining nodes assigned to gaps. If such an upper limit is not set, SEGA calculates a global graph alignment, considering all nodes in a graph. A limit below c_{max} can be regarded as a stringency constraint for the partial alignment that controls the tolerated amount of structural deviation. From a biological point of view, it might be reasonable to set such an upper limit and retrieve just a partial alignment, e.g., for two binding sites that share a similar subpocket while being globally dissimilar. Of course, the choice of a proper threshold is not clear in advance and should be chosen based on the application at hand. The complete procedure is summarized in pseudo-code in Algorithm 2.

Algorithm 2 SEGA: Constructs a global alignment A for the graphs G_1, G_2

Require: distance matrix D, graph $G_1 = \{V_1, E_1\}$, Graph $G_2 = \{V_2, E_2\}$
  $S = \emptyset, W_1 = \emptyset, W_2 = \emptyset$
  $k = 0$
  while $c_k \leq c_{max}$ **do**
    for all $(v_i^{(1)}, v_j^{(2)})$ with $d_{ij} \leq c_k$ **do**
      $f_{c_k}(v_i^{(1)}) \leftarrow \{v_j^{(2)} \in V_2 \,|\, d_{ij} \leq c_k\}$
      $g_{c_k}(v_j^{(2)}) \leftarrow \{v_i^{(1)} \in V_1 \,|\, d_{ij} \leq c_k\}$
      for all $(v_i^{(1)}, v_j^{(2)})$ **do**
        if $f_{c_k}(v_i^{(1)}) = \{v_j^{(2)}\}$ and $g_{c_k}(v_j^{(2)}) = \{v_i^{(1)}\}$ **then**
          add $(v_i^{(1)}, v_j^{(2)})$ to A, add $v_i^{(1)}$ to W_1, add $v_j^{(2)}$ to W_2
        else
          add $(v_i^{(1)}, v_j^{(2)})$ to S
      if $|A| > 4$ **then**
        if $S \neq \emptyset$ **then**
          $U_1^k \leftarrow \{v_i^{(1)} \in V_1 \,|\, f_{c_k}(v_i^{(1)}) \neq \emptyset\} \setminus W_1$
          $U_2^k \leftarrow \{v_j^{(2)} \in V_2 \,|\, g_{c_k}(v_j^{(2)}) \neq \emptyset\} \setminus W_2$
          $M, C \leftarrow construct\_matrix(U_1^k, U_2^k, S, D)$
          $A_H \leftarrow hungarian\_algorithm(M)$
          for all $(v_i^{(1)}, v_j^{(2)}) \in A_H, M_{ij} < C$ **do**
            add $(v_i^{(1)}, v_j^{(2)})$ to A, add $v_i^{(1)}$ to W_1, add $v_j^{(2)}$ to W_2
            $S \leftarrow \emptyset$
      else
        $S \leftarrow S \cup A, A \leftarrow \emptyset, W_1 \leftarrow \emptyset, W_2 \leftarrow \emptyset$
        $S_4 = \{X \subset S \,|\, |X| = 4\}$
        if $S_4 \neq \emptyset$ **then**
          $S_{min} \leftarrow X \subset S_4, dev(X) \leq dev(X'), X' \subset S_4$ {select S_{min} with minimal spatial deviation $dev(S_{min})$}
          for all $(v_i^{(1)}, v_j^{(2)}) \in S_m$ **do**
            add $(v_i^{(1)}, v_j^{(2)})$ to A, add $v_i^{(1)}$ to W_1, add $v_j^{(2)}$ to W_2
      $k = k + 1$
  return Alignment A for the graphs G_1, G_2

Algorithm 3 construct_matrix: Constructs a cost matrix from ambiguous mapping candidates

Require: S set of candidate mappings of nodes, U_1^k, U_2^k
  $C = Max\_Value$
  for all $v_i^{(1)} \in U_1^k$ **do**
    for all $v_j^{(2)} \in U_2^k$ **do**
      if $(v_i^{(1)}, v_j^{(2)}) \notin S$ **then**
        $M_{ij} = C$
      else
        $M_{ij} = \sum_{q=1,2...|W_1|} \left| |v_i^{(1)} - w_q^{(1)}| - |v_j^{(2)} - w_q^{(2)}| \right|,$
        $w_q^{(1)} \in W_1, w_q^{(2)} \in W_2$
  return cost matrix M, constant C

While theoretically more useful, the question remains whether the above suggested strategy will yield an improvement over simply generating the alignment by solving a weighted optimal assignment problem. Thus, as an alternative, the SEGAHA (SEmi-global Graph ALignment - Hungarian Algorithm) variant is proposed as an alternative, where the incremental assignment of nodes is replaced by the Hungarian algorithm (Kuhn, 2005). The question, which of the two approaches will be more suitable for protein binding site comparison will be addressed in Chapter 5.

4.3.1.3 Defining a distance measure

If not specified otherwise, the above-presented SEGA algorithm produces a global graph alignment, deriving an assignment between all constituents of a binding site. Binding sites might share several common subpockets that are not necessarily arranged in the same manner. Since SEGA ideally will only rarely resort to global information, such subpockets should still be mutually assigned by the procedure. However, this means that quality measures based on root mean squared deviation (RMSD) that are usually applied in more strict procedures, e.g., template-based approaches (Barker and Thornton, 2003; Stark and Russell, 2003), are not applicable, except for cases where the binding sites are globally similar. Thus, a more general quality measure is needed to rate the alignment quality.

To define a more general, size-independent quality measure of the quality of the alignment A, that can be interpreted as distance between the two structures G_1 and G_2, one can proceed from a measure that can be seen as a degree of inclusion of G_1 in G_2:

$$\delta(G_1,G_2) = \frac{\sum_{(v_i^{(1)},v_j^{(2)})\in A} d_{ij} + c_p \cdot (|A| - |G_1|)}{|G_1|} \ . \tag{4.30}$$

The constant c_p is a penalty that accounts for unmatched nodes which can simply be set to the highest obtainable distance if no triangles can be matched. A degree of inclusion of G_2 in G_1 is defined analogously.

Based on (4.30), two measures of distance between G_1 and G_2 can be defined, a "conjunctive" and a "disjunctive" one, in analogy to the scoring scheme for the GAVEO approach:

$$\Delta_{max}(G_1,G_2) = \max\{\delta(G_1,G_2),\delta(G_2,G_1)\} \tag{4.31}$$

$$\Delta_{min}(G_1,G_2) = \min\{\delta(G_1,G_2),\delta(G_2,G_1)\} \tag{4.32}$$

In this case, the measure (4.31) can be seen as a relaxed equality in terms of two-sided inclusion ($A \subset B$ and $B \subset A$), while the disjunctive combination, favoring a one-sided inclusion, is given by (4.32). Again,

$$\Delta_{min}(G_1,G_2) \leq \Delta_{max}(G_1,G_2) \ .$$

The question which of these two measures, the conjunctive or the disjunctive one, yields more suitable degrees of similarity cannot be answered in general and instead depends on the problem setting, in particular on the purpose for which the similarity is used (e.g., function prediction) and the way in which protein binding sites are extracted and modeled (e.g., whether or not the model may include parts of the protein not belonging to the binding site itself).

Again, to account for both possible extremes while allowing for a certain degree of flexibility, the ultimate distance measure of the SEGA algorithm is defined as a (linear) combination of (4.31) and (4.32):

$$\Delta(G_1,G_2) = \alpha \cdot \Delta_{max}(G_1,G_2) + (1-\alpha) \cdot \Delta_{min}(G_1,G_2) \ . \tag{4.33}$$

This distance is again inversely related to a similarity score.

Similar to GAVEO, (4.33) is a special case of an OWA (ordered weighted average) aggregation of the two degrees of inclusion, G_1 and G_2, and the parameter $\alpha \in [0,1]$

controls the trade-off between the two extreme aggregation modes: The closer α is to 1, the closer the aggregation is to the minimum, i.e., the more demanding it becomes. The value α corresponds to the "degree of andness" of the aggregation (4.33), i.e., the degree to which this aggregation behaves like a conjunctive combination (Fodor and Roubens, 1994); likewise, $1 - \alpha$ corresponds to the "degree of orness".

In principle, choosing a high value of α favors the detection of largely similar binding sites, thus yielding results more alike to those of global methods. This can be of interest for proteins belonging to the same protein family or fold. A low value of α would be beneficial for the detection of more remote similarities, which could be more useful to detect similarities in proteins of different folds.

5

Results and Discussion

In the following, the approaches introduced in the previous chapters will be experimentally validated and compared on different datasets as well as different problem settings. The following algorithms will be used in the experiments:

- BFPH (Bin-FingerPrints Hamming): Crisp fingerprints using a binning of the edge lengths with bin size b. Fingerprints are compared using the Hamming distance (4.2.2.1).
- BFPJ (Bin-FingerPrints Jaccard): Crisp fingerprints using a binning of the edge lengths with bin size b. Fingerprints are compared using the Jaccard distance (4.2.2.1).
- GAVEO (Graph Alignments Via Evolutionary Optimization): Evolutionary algorithm that optimizes an objective function based on a graph edit distance (4.1.1).
- GAVEO* : GAVEO in combination with the scoring function originally used by the greedy heuristic (4.1.1).
- GAVEOc (GAVEO with preserved clique): A variant of GAVEO where the maximal clique is calculated prior to the optimization and preserved throughout the calculation (4.1.1.4).
- FPH (ε-FingerPrints Hamming): Crisp fingerprints using a fixed set of edge labels $l \in \{1, ..., 12\}$ in conjunction with an ε threshold. Fingerprints are compared using the Hamming distance (4.2.2.1).

- FPJ (ε-FingerPrints Jaccard): Crisp fingerprints using a fixed set of edge labels $l \in \{1, ..., 12\}$ in conjunction with an ε threshold. Fingerprints are compared using the Jaccard distance (4.2.2.1).

- FFP (Fuzzy FingerPrints): Fuzzy fingerprints using triangular membership functions controlled by the radius parameter η. Fingerprints are compared using a generalization of the Jaccard measure (4.2.2.2).

- SEGA (SEmi-global Graph Alignment): A semi-global approach using local similarities and global information to construct a global graph alignment from a distance matrix D (4.3.1).

- SEGAHA (SEGA with Hungarian Algorithm): A variant of SEGA, where the Hungarian algorithm (Kuhn, 2005) is used to construct a global alignment from the distance matrix D by calculating a cost-minimal assignment of nodes (4.3.1).

To better judge the performance of the introduced algorithms, some baseline algorithms will be employed, using sequence information as well as structural information retrieved from CavBase.

- BK (Bron-Kerbosch algorithm): A clique-enumeration algorithm (Bron and Kerbosch, 1973) commonly used in graph-based protein structure comparison ((Kinoshita and Nakamura, 2005; Redfern et al., 2007; Schmitt et al., 2001), cf. Chapter 2). The Bron-Kerbosch is used in CavBase to calculate the first 100 cliques instead of a full enumeration which proved already sufficient to create meaningful solutions (Schmitt et al., 2002). Therefore, the Bron-Kerbosch algorithm will be used analogously here.

- CB (CavBase clique algorithm): The original algorithm used in CavBase, which represents a combination of the Bron-Kerbosch approach with a surface-based scoring scheme (Schmitt et al., 2001).

- GH (Greedy Heuristic): A greedy heuristic based on clique detection. This approach was developed in a previous work and represents the most recent approach for the comparison of CavBase data (Weskamp, 2007).

- RW (Random Walk kernel): The random walk kernel of Gärtner (2003) with the extensions introduced in Chapter 4 (4.2.1.1).

- SA (Sequence Alignment): A local sequence alignment using the Smith-Waterman algorithm (Smith and Waterman, 1981) as implemented in the jaligner tool (Moustafa, 2005).
- SP (Shortest Path kernel): The shortest path kernel of Borgwardt and Kriegel (2005) with the extensions introduced in Chapter 4 (4.2.1.2).
- SPSA (SP with Sequence Alignment): The shortest path kernel expansion using sequence alignment on paths (4.2.1.2).

Although the random walk and the shortest path kernels were expanded during this thesis to be applicable on the protein binding site model, they were originally suggested elsewhere. Hence it is more appropriate to regard them as baseline approaches for the local comparison approaches. In the experiments, the kernel measures are used directly as similarity measures [1].

Note that some of the approaches might fail to calculate comparisons for pairs of exceedingly large graphs. This especially pertains to the CavBase approach and the random walk kernel, to a lesser extend also to BK and GH. In these cases, the corresponding score is set to $-\infty$ in case of similarity scores, respectively to ∞ in case of distance values.

This chapter is organized as follows: First, an overview of the datasets used in the experiments is given prior to the actual experimental part. The experimental part starts with several preliminary experiments aiming at deriving suitable parameter settings for the different approaches before more time-consuming studies are conducted.

This is followed by an assessment of the algorithmic performance of the different approaches when confronted with different levels of structural and mutational distortion. Section 5.7 presents a number of classification experiments on different datasets used to assess the performance and suitability of the presented methods for classification tasks, as the main goal of the graph comparison algorithms presented in this thesis is to discriminate between different classes of protein binding sites. Section 5.6 presents results for another typical application of protein structure comparison tools, the retrieval of similar structures from a reference dataset.

In Section 5.8, the suitability of the presented algorithms for comparison tasks beyond experimentally derived structures and protein binding sites is addressed.

[1] In preliminary experiments (not shown), the use of the kernel distance (4.10) showed no improvement of classification results over using the kernel measure directly

Abbr.	Algorithm
BFPH	Bin-FingerPrints using the Hamming distance
BFPJ	Bin-FingerPrints using the Jaccard coefficient
BK	Bron-Kerbosch algorithm
CB	CavBase approach
GAVEO	Graph Alignment Via Evolutionary Optimization
GAVEO*	GAVEO + original similarity measure
GAVEOc	GAVEO + preserved clique
GH	Greedy Heuristic
FPH	ε-FingerPrints using the Hamming distance
FPJ	ε-Fingerprints using the Jaccard coefficient
FFP	Fuzzy Fingerprints
RW	Random Walk kernel
SA	Sequence Alignment (Smith-Watermann)
SEGA	SEmi-global Graph Alignment
SEGAHA	SEGA using the Hungarian Algorithm
SP	Shortest Path kernel
SPSA	Shortest Path kernel with Sequence Alignment

Table 5.1: Algorithms used during the experiments.

5.1 Datasets

The main focus of this thesis is on the comparison of protein binding sites according to the pseudocenter model introduced in Chapter 3. Therefore, all binding sites where retrieved from the CavBase database Schmitt et al. (2002) which is a part of the ReliBase+ database hosted by the Cambridge Crystallographic Data Center (CCDC) (Hendlich et al., 2003). Currently, CavBase contains 308,141 cavities extracted from 70,850 PDB entries (June 2011).

From this database, several smaller datasets where constructed to assess the performance of the different algorithms, with different objectives in mind. This was necessary, since an all-against-all comparison of all these structures is infeasible without the use of high-performance computing facilities.

As one major application for the developed approaches is the classification and comparison of protein binding sites with respect to the accommodated ligand, an initial benchmark dataset was constructed by drawing from the two most highly populated groups of binding sites in the CavBase database. Thus, a two-class classification dataset was constructed, which contained protein binding sites known to host either adenosine-5'-triphosphate (ATP) or nicotinamide adenine dinucleotide (NADH). Both molecules act as cofactors for a plethora of functionally and phylogenetically diverse proteins and can bind to the proteins in different conformations. Hence, two randomly drawn cavities, even if hosting the same ligand, do not necessarily share a common geometric architecture. Therefore, a subset of these two groups was selected.

The main purpose of this initial dataset was to assess the classification performance of the previously introduced approaches, especially the global methods. One the one hand, this demands a dataset for which binding pockets of the same class show some structural resemblance, which is obviously not the case for the complete set of ATP or NADH binding pockets. On the other hand, since all approaches were developed to tolerate structural variance to a certain extend, the binding pockets should also not be too similar. Hence, instead of simply selecting proteins based on sequence similarity, cavities were instead selected by drawing on ligand information.

More precisely, ligands were superimposed using Kabsch's algorithm (Kabsch, 1976) and then clustered according to the root mean squared deviation (RMSD) of the superimposed molecules. Subsequently, subsets were selected, for which the difference in RMSD ranged up to 0.4 Å. Thereby it is assured that the ligands are at least bound in similar conformation although not necessarily orientation. The RMSD difference threshold can be regarded as a compromise between conformational similarity of the ligand and dataset size. At the given threshold, a sufficiently large dataset (ATP/NADH dataset) for an all against all comparison was obtained, containing 355 protein binding sites in total, 214 NADH binding sites and 141 ATP binding sites. To keep the runtime requirements low, no further cavities were included, although the threshold is well below the accepted RMSD difference of a successful docking solution, which would be 2 Å (Verdonk et al., 2008).

While this first dataset ensures a certain similarity of the ligand conformations by enriching structurally similar binding sites, it also favors binding sites belonging to proteins that are related on the sequence and/or fold level. Indeed, the dataset contained only 15 different folds. A main motivation for the development of the presented approaches was to uncover non-trivial similarities, i.e., similarities not apparent on the sequence or fold level.

Hence, a more challenging ATP-NADH dataset (subsequently termed 1-fold ATP/NADH dataset) was created, including only remotely similar binding sites that belong to different folds according to the SCOP (Structural Classification Of Proteins) database (Murzin et al., 1995). Proteins taken from the complete set of ATP and NADH binding proteins stored in CavBase were filtered to include only one protein per SCOP fold, creating a non-redundant dataset. The resulting dataset contained only binding sites of proteins, that do not exhibit similarity on the folding level, although a structural similarity between the binding sites themselves might still exist, based on the notion that proteins with different folds can still accommodate similar ligands. The dataset was constructed to assess, whether a comparison of the protein binding sites alone can still retrieve similarities, even if none are apparent on the folding level and therefore the global protein structure is different.

To obtain a suitable parameter setting for the subsequent experiments, another four class dataset was constructed by randomly drawing 50 cavities per class from all cavities containing either ATP, NADH or FAD (flavin adenine dinucleotide), three of the most abundant ligands in the CavBase. To include also a more rigid group of structures, a fourth class corresponding to cavities containing a porphyrine ring as a ligand was included.

In the retrieval experiments, a high-resolution subset of the CavBase database was used to keep the runtime requirements at a manageable level. This was necessary due to the comparably high runtime requirements of some some of the algorithms. By including only cavities derived from protein complexes with a minimal resolution of 2.5 Å in the high-resolution subset, the number of necessary comparisons could be reduced by one-third compared to using the complete set of cavities in CavBase. The final dataset (HiRes) contained 186,507 cavities but still denoted a representative subset of the complete CavBase.

As a further external benchmark set, a set of representative proteins constructed for the evaluation of SiteEngine was included in the experiments. SiteEngine, as mentioned in Chapter 2, is another surface-based protein comparison approach which operates on a concept of binding pockets similar to CavBase. This dataset was originally compiled to include several structurally different classes of proteins, among them fatty acid-binding proteins, serine proteases, adenine-containing ligands and others (for a detailed description, see (Shulman-Peleg et al., 2004)). A summary of the dataset, including the classes defined by Shulman-Peleg *et al.* and the corresponding PDB codes can be found in Table A.1. Since the SiteEngine model differs from the CavBase model in some aspects, especially modeling and extraction of the binding sites, the PDB codes where used to extract the associated cavities from CavBase to construct a set of corresponding cavities.

Another externally compiled benchmark dataset used in retrieval and classification experiments is the Astex Non-native Set (Verdonk et al., 2008), initially constructed for the assessment of the performance of docking algorithms. The dataset was based on the Astex Diverse Set Hartshorn et al. (2007), another docking benchmark dataset containing

85 different high-quality protein-ligand complexes. In this dataset, each ligand is represented once. The Astex Non-native Set contains different conformations of the same protein targets that are addressed by the ligand (i.e., either apo structures or structures complexed with different ligands), thus being a more realistic and challenging docking benchmark dataset. The resulting benchmark set allows to assess the performance of the different methods when confronted with structural variation mainly due to ligand-induced protein conformation changes. Only structures with a minimal resolution of 2.5 Å were included. After filtering and visual inspection, the Astex Non-native Set contained 1112 non-native structures for 65 of the original 85 ligands.

Additionally, to apply the approaches to a problem beyond the realm of binding sites, an HIV mutant sequence dataset was used. The objective here was to distinguish different HIV mutants. The sequence dataset was compiled from the HIV sequence database at Los Alamos National Laboratory in a previous study by Sander et al. (2007), which contained 1100 sequences of the HIV glycoprotein 120 (gp120) derived from clonal samples of 332 patients. By discarding duplicate sequences, a set of 514 mutant variants of the V3 loop of gp120 were derived, a region of the protein which plays an important role in the cellular adhesion process and the infiltration of the host cell. Each of the mutant strains are annotated as being either of the X4, R5/X4 or the R5 phenotype, indicating their capability of interacting with two different chemokine receptors, CCR5 and CXCR4, thus playing a dominant role in the progression of the AIDS disease. To obtain a structural representation of the mutants, the protein structure of the HIV-1 JR-FL gp120 from the PDB (PDB code: 2b4c (Huang et al., 2005)) was used as a template to obtain a structure prediction with RCSB, a structure prediction algorithm using threading (Canutescu et al., 2003). Sequences were aligned with MUSCLE (Edgar, 2004) and the V3 loop ranging from residues 296 to 331 of PDB structure 2b4c was used as a backbone template for the modeling of the mutant sequences. Of the 514 mutants 82 contained insertions or deletions compared to the template sequence, which is likely to reduce the quality of the structure prediction. Hence they where excluded from the dataset, yielding 432 mutually distinct sequences. Subsequently, the predicted structures were transformed to a pseudocenter representation according to the CavBase rules as introduced in (Kuhn et al., 2006; Schmitt et al., 2002).

None of these datasets where used as test datasets during the development of the approaches, with the only exception of the ATP/NADH dataset.

5.2 Parameter influence on algorithmic performance

5.2.1 Empirical estimation of parameter settings

As all of the approaches presented in Chapter 4 are parameterized, some preliminary experiments were conducted to assess the parameter influence on the performance of the developed algorithms and to obtain a suitable parameter setting for further experiments. As each of the approaches is founded on a unique strategy, these experiments had to be tailored towards the specific needs of the algorithms.

For the GAVEO and GAVEOc approaches, the evolutionary optimization is controlled by several parameters that mainly influence alignment quality and runtime performance. In these two cases, parameters were estimated using sequential parameter optimization. In case of the other approaches, each method is parametrized by a single parameter. In order to find a suitable setting for this parameter, the four-class dataset introduced in the previous section will be employed as a test dataset in a classification scenario.

In the following, the global GAVEO and GAVEOc approaches will be considered first, followed by the more simple local fingerprint approaches. Finally, the semi-global SEGA algorithm is addressed.

5.2.1.1 Sequential optimization of GAVEO parameters

As outlined in Chapter 4, the evolutionary optimization of graph alignments via GAVEO is controlled by several parameters, influencing the evolutionary operators as well as termination criteria. An unfavorable setting of these parameters might result in near-random alignments of low quality, hence a reasonable setting of parameters is vital for the performance of the algorithm. The sequential parameter optimization toolbox (SPOT) developed by Bartz-Beielstein (2006) presents a convenient tool to derive such a setting for multiple parameters simultaneously.

Since the GAVEO approach uses the same scoring function as the greedy heuristic, the scoring parameters controlling mismatch penalties and match rewards originally suggested by Weskamp (2007) will be used. Instead, the focus of this experiment will be on the additional parameters unique to GAVEO, namely μ, ν, ρ, p_{check} and self-adaption.

The result of the evolutionary optimization will strongly depend on the invested time, yet a low runtime is preferable with respect to efficiency. Therefore, the goal of this experiment was to obtain a suitable parameter setting for the exogenous parameters in order to minimize the runtime needed to arrive at an optimal or near-optimal solution. Obviously,

parameter	μ	ν	ρ	p_{check}	self-adaption	stall generations
range	$[1,20]_{\mathbb{N}}$	$[0.1,30]$	$[1,2]_{\mathbb{N}}$	$[0,1]$	$[0,1]_{\mathbb{N}}$	$[10,500]_{\mathbb{N}}$
estimate	4	6	2	0.5	0	200

Table 5.2: Parameter setting for the GAVEO and GAVEOc approach

this means that the optimal solution must be known in advance. For this reason, only identical binding sites where aligned for which the optimal solution is trivially known.

In the SPOT experiment, 20 binding sites randomly drawn from CavBase where aligned to themselves, starting at randomly initialized alignments. As the GAVEO approach will mainly be used for pairwise comparisons, the SPOT experiment is carried out for the pairwise case. Parameters where automatically adjusted using the SPOT toolbox. SPOT creates a number of so-called design points in parameter space used to estimate a setting that minimizes the calculation time necessary to arrive at the optimal solution. The elapsed runtime was used as a fitness criterion. Since only correct solutions should be considered, the fitness value was set to infinity, if the optimal solution was not found.

In addition to the exogenous parameters, the termination criterion is crucial for the quality of the results. In order to achieve a possibly high quality of the results, it is important to avoid limiting the algorithm too much by setting too stringent termination conditions. Therefore, only the number of stall generations was used as termination criterion. On the other hand, a low runtime investment is preferred, hence, the number of stall generations should be set as small as possible. To obtain an estimate on the minimum number of stall generations required to achieve at a high quality solution, stall generations where also included as a parameter in the SPOT experiment. Again, the fitness value was set to ∞, if the optimal solution was not obtained.

In total, 6 parameters where investigated: the population size μ, the selective pressure ν, the recombination parameter ρ, the check probability for gap columns p_{check}, the number of stall generations and the self adaption of the mutation step size (which can be switched on or off). The obtained parameter setting is given in Table 5.2. Fig. 5.1 shows the fitness landscapes for pairwise combinations of parameters.

As can be seen in Fig. 5.1d and Fig. 5.1e ($\rho = 2$), allowing recombination is preferable, even in the pairwise case. Self-adaption instead is set to 0, indicating that an adaption of the mutation strength is counterproductive. The check probability p_{check} should be high, while the population parameters ν and μ can apparently be set to relatively low values. An interesting result is the estimation of the stall generations. As can be seen

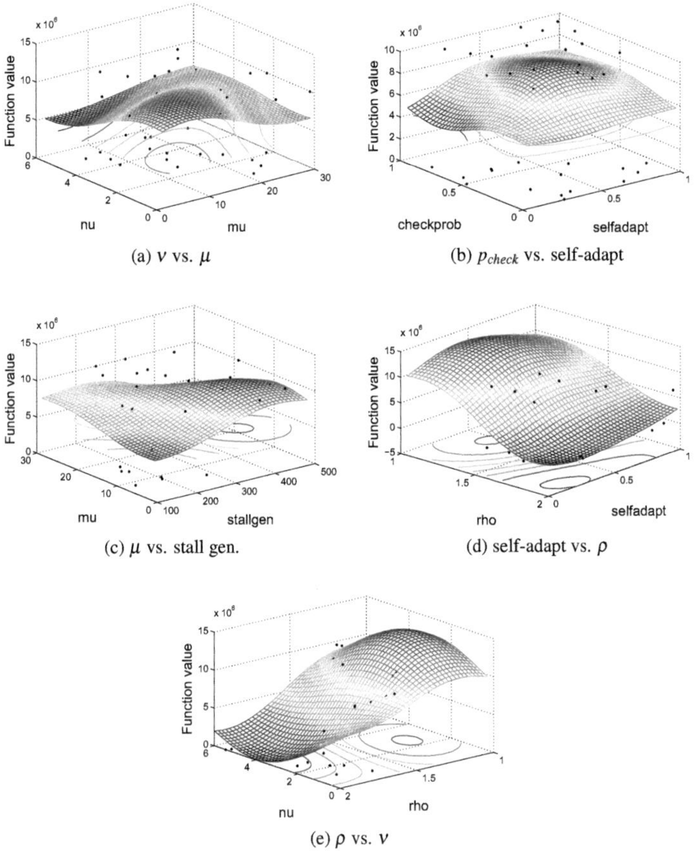

(a) ν vs. μ

(b) p_{check} vs. self-adapt

(c) μ vs. stall gen.

(d) self-adapt vs. ρ

(e) ρ vs. ν

Figure 5.1: Fitness landscapes for different parameter combinations in the SPOT experiment. A higher function value corresponds to higher runtimes.

in Fig. 5.1c, GAVEO is able to find the optimal solution with less than 200 stall generations. The number of stall generations was estimated to 200 by SPOT, increasing this number had no further benefit. In the following experiments, the parameters will be set as indicated in Table 5.2.

One might argue, that this experiment is a bit extreme, given that only a global optimal solution is accepted. For the purpose of classification, a near-optimal solution might suffice, which could be derived with fewer stall generations. Yet, it is hard to judge at which point the solution is sufficiently optimized. Since the main goal of this experiment is to find a reasonable parameter setting, it is arguably better to play safe than to risk a diminished performance due to a premature termination of the optimization. Moreover, for interpreting the alignments, an optimal solution should be preferred.

For the GAVEOc variant, repeating the SPOT experiment would be meaningless, since the precalculated maximal clique solution should already retrieve the correct solution without any need for optimization. However, since the evolutionary optimization process is identical to the standard GAVEO approach, the same settings will be employed for all experiments conducted with GAVEOc.

5.2.1.2 Influence of tolerance parameters on fingerprints performance

Regardless of the used fingerprint variant, the algorithmic performance will obviously depend on the defined patterns and therefore be influenced by the parametrization used to construct the fingerprint vectors. For each of the fingerprint approaches, this relates to the way edge length tolerance is realized. In each case this is controlled by a single parameter.

In the case of using crisp fingerprints, the fingerprint construction mainly depends on either a tolerance threshold ε or the bin size b, if a binning strategy is employed. For the fuzzy fingerprints, this is controlled by the parameter η, that controls the support of the triangular fuzzy sets used to model the fuzzy edge labels of the patterns.

The experimental setup of the GAVEO approach is useless here, as the comparison of identical binding sites will always yield the highest possible score, regardless of the parametrization. Instead, the influence of the parameter setting on the performance will be assessed in a classification scenario. More precisely, classification experiments were conducted on the four class test dataset described in Section 5.1, consisting of four classes of protein binding sites, as defined by the ligand co-crystallized with the protein structure: ATP, NADH, FAD and a porphyrine ring. The first three classes represent three of the most highly populated ligand classes in the CavBase. Additionally, since these ligands are relatively flexible and bound by many diverse proteins, another group containing a more rigid ligand in the form of heme and related porphyrine ring structures.

Algorithmic performance was assessed in terms of misclassification rate of the four-class problem, using ten-fold stratified cross validation in conjunction with k-nearest neighbor classification.

Crisp fingerprints with tolerance threshold

When realizing crisps fingerprints with a tolerance threshold ε, two edge weights will be considered matching, if their absolute difference is equal to or lower than ε. Hence, the higher ε, the more tolerant the matching becomes, allowing for structural variation, but also increasing the risk of becoming less discriminative. It is difficult to estimate in advance, how this trade-off between tolerance and specificity should be realized.

Thus, classification experiments where performed for different settings of ε on the above mentioned four-class problem. Algorithmic performance was assessed in terms of classification accuracy. Fig. 5.2 depicts the obtained results for $k = 1$, which generally yielded the best performance, regardless of ε. The complete results can be found in Table B.2 in Appendix B.

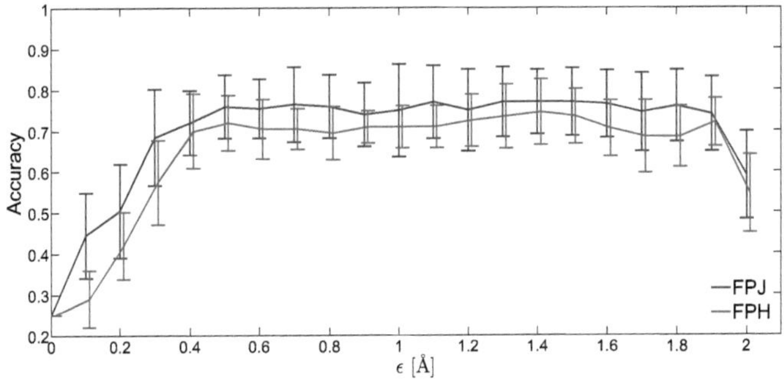

Figure 5.2: Results of a ten-fold stratified cross-validation using a 1-nearest neighbor classification based on the FP approach. Classification accuracy on the four-class dataset is plotted for different threshold values ε.

Setting ε too low obviously has a detrimental effect, regardless of the comparison measure, which is not surprising, since some tolerance is needed to account for minor structural differences that even exist in different crystal structures of identical proteins. For $0.5 \leq \varepsilon \leq 1.5$, performance changes only marginally, indicating that the exact choice of epsilon is not crucial here. At this point, it should be mention that a tolerance of up to 2 Å is not unusual for structural comparison tools and many approaches are known to tolerate such differences (Orengo and Taylor, 1996; Schmitt et al., 2002), thus the parameter range is still in a reasonable. For $\varepsilon > 1.5$, the performance starts to decrease again, indicating the loss of geometric information by becoming too error-tolerant.

The best performance is obtained for $\varepsilon = 0.5$, although this might vary for different datasets, especially since there is not much difference to higher values of ε. However, it seems clear that ε should not be set below 0.5. At this threshold, the crisp fingerprints also achieve a complete coverage of the domain of edge weights up to the upper limit of 12 Å gaps, whereas for lower values some edge lengths will be lost. Increasing ε beyond that 0.5 Å would mean that edge weights would be assigned to different integer labels simultaneously. Hence, in further experiments, ε will be set to 0.5.

Crisp fingerprints with binning

If a binning strategy is used, the bin size b will control which edge weights are considered matches, respectively mismatches. The smaller b is set, the more fine-grained the separation will become, similar to setting a low ε-threshold. Yet it would not be entirely correct to regard b as a tolerance parameter, since edge weights that differ by a fraction of b might still be considered as mismatches.

Again, the classification accuracy was obtained for different settings of b by ten-fold stratified cross validation. Fig. 5.3 depicts the obtained results. The complete results can be found in Table B.1 in Appendix B.

Similar to the results obtained for the ε-based approach, setting b too low should be avoided. For higher values ($b > 5$), the accuracy is decreasing rapidly, where the difference seems again to be marginal for medium b. The best results are obtained for $b = 2.5$ in case of the Hamming distance. For the Jaccard coefficient, $b = 3$ yields slightly better results. Again, given the small differences in accuracy, this might be different for other data sets. However, the choice of b appears to be less critical as expected. In further experiments, the parameterization $b = 2.5$ will be realized. This might seem large at first, but as mentioned above, a tolerance of 2 Å is not uncommon.

Influence of fingerprint construction on classification performance

Finally, the performance of the fuzzy fingerprints will depend on the choice of membership functions used to construct the fingerprints. The fuzzy fingerprints as introduced

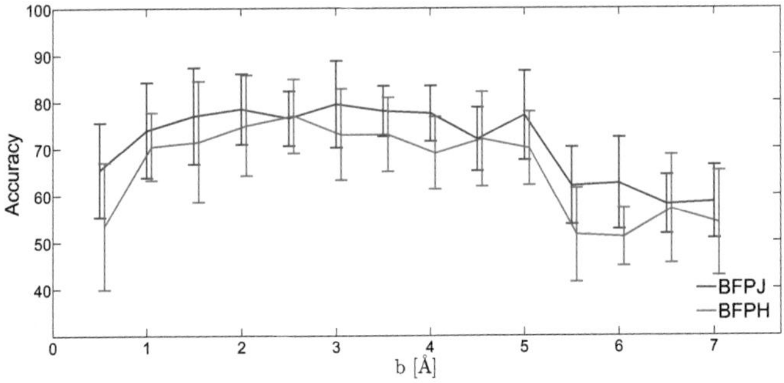

Figure 5.3: Results of a ten-fold stratified cross-validation using a 1-nearest neighbor classification based on the BFP approach. Classification accuracy on the four-class dataset is plotted for different bin sizes *b*.

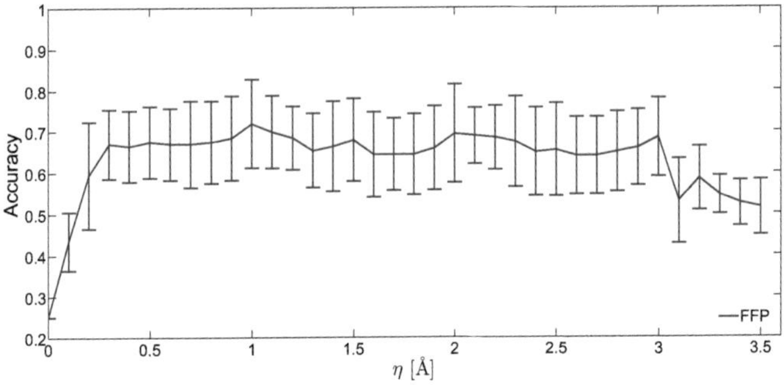

Figure 5.4: Results of a ten-fold stratified cross-validation using a k-nearest neighbor classification based on the FFP approach. Classification accuracy is plotted for different η.

in Chapter 4 are restricted to triangular membership functions, whose support (or more intuitively, the width of the triangles) is controlled by η.

Similar to ε, η can be regarded as a tolerance parameter: the larger the support of a membership function, the more strongly can an edge weight deviate from the core of the fuzzy set and still contribute to the fuzzy label. To get an impression of the influence of η on the algorithmic performance, classification experiments were once again conducted via ten-fold stratified cross validation. The results for $k = 1$ are displayed in Fig. 5.4.

As was the case in the previous experiments, setting the parameter too low, and thus becoming to stringent, should be avoided. In this case, the results show that η should not be below 0.3 Å. The best results are obtained for $\eta = 1$, although the performance does not change as dramatically for $0.3 \leq \eta \leq 3$, similar to the other fingerprint variants. for $\eta > 3$, performance starts to deteriorate again. In further experiments, η will be set to 1.

These three experiments also allow a first comparison of the different fingerprint approaches. Table 5.3 shows the classification accuracy for the different fingerprint variants.

Although all fingerprint methods show comparable performance, using fingerprints based on a binning strategy outperforms both fuzzy fingerprints and fingerprints using an ε threshold, regardless of the similarity measure, at least on the four-class problem. Further experiments will reveal whether this is a random effect or a genuine superiority of the binning variant.

Moreover, the Jaccard measure generally yields better results than the Hamming distance. Again, whether this holds true for other problems as well remains to be examined. Note that the parameter setting has no influence on the runtime complexity: regardless of the number of patterns, each subgraph of size three has to be considered once. This is different for the SEGA algorithm.

k	BFPH	BFPJ	FPH	FPJ	FFP
1	77.5	**80.0**	72.0	76.0	73.0
3	75.5	73.0	69.0	67.5	63.5
5	68.5	71.0	60.0	60.0	59.5
7	66.0	69.5	56.0	58.5	57.0
9	68.0	68.5	51.5	53.5	56.5

Table 5.3: Classification accuracy for the different fingerprint approaches derived by 10-fold stratified cross validation using a k-nearest neighbor classifier.

5.2.1.3 Influence of the neighborhood parameter on the performance of SEGA

The performance of the SEGA algorithm will obviously depend on the choice of the neighborhood size, i.e., the neighborhood parameter n_{neigh}. Choosing a suitable setting for n_{neigh} is not trivial in advance. On the one hand, the higher n_{neigh}, the more information of the immediate surrounding of a pseudocenter is considered, thus it could be argued that the neighborhood parameter should be set to a high value. On the other hand, with increasing neighborhood SEGA will compare substructures of increasing size, up to an extreme where the whole graph will be considered. Thus, the SEGA method "converges"

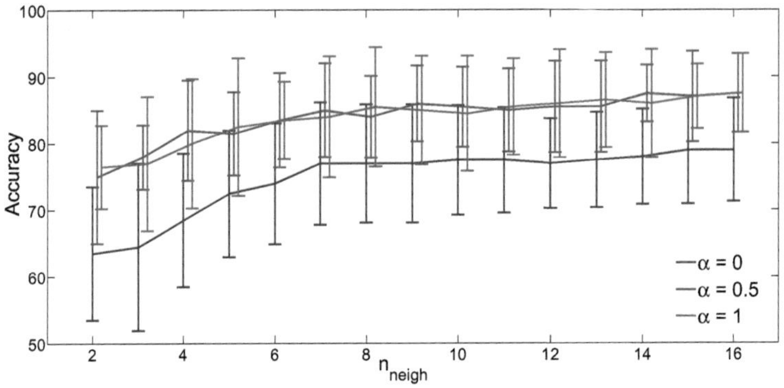

Figure 5.5: Performance of 1-nearest neighbor classification in a ten-fold stratified cross validation on the four class dataset. The misclassification rate is plotted for different values of n_{neigh}.

to a global method with increasing size, loosing the merits of a semi-global approach, such as flexibility. Moreover, runtime requirements will increase with the n_{neigh}, thus a low neighborhood size is desirable.

To estimate the optimal setting for n_{neigh}, classification experiments were again conducted as above, realizing a ten-fold stratified cross validation. Classification was done with a k-nearest neighbor classifier based on the distance score produced by the SEGA method. The results were used to assess the influence of n_{neigh} on prediction accuracy and runtime requirements. Note, that unlike the fingerprint approaches, the neighborhood parameter will directly influence the runtime requirements of SEGA, as larger neighborhood graphs will be more time-consuming to compare.

Fig. 5.5 depicts the classification results for different values of n_{neigh} with a fixed α. Results are depicted for $k = 1$.

As outlined in Chapter 4, the scoring functions of both the global GAVEO approach and the semi-global SEGA algorithm are realized as a combination of a conjunctive measure, favoring a mutual inclusion of graphs for a high similarity score and a disjunctive measure, focusing on one-sided inclusion. The trade-off between both extremes is controlled by the parameter α [2].

[2] In case of SEGA, a high α emphasizes mutual inclusion.

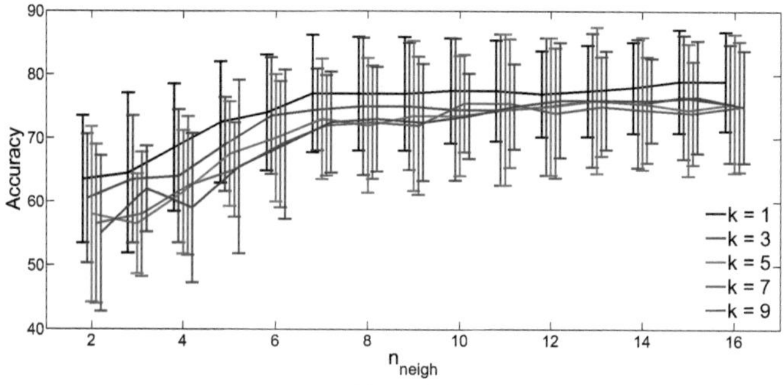

(a) $\alpha = 0$

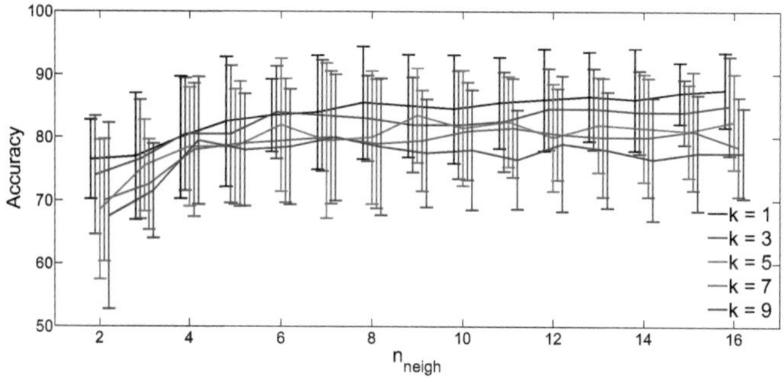

(b) $\alpha = 1$

Figure 5.6: Performance of k-nearest neighbor classification in a ten-fold stratified cross validation on the four class dataset for $\alpha = 1$ and $\alpha = 0$. The misclassification rate is plotted for different values of n_{neigh}.

The results for $\alpha = 0$ and $\alpha = 1$ are shown again for different k in Fig. 5.6a and Fig. 5.6b. Again, the complete results can be found in the Appendix B, Table B.4.

In each case, the classification accuracy increases with n_{neigh}, as was to be expected, since increasing n_{neigh} means considering more neighborhood information. Accuracy remains comparably stable, once a certain neighborhood size is reached, increasing only slightly for $n_{neigh} > 8$. This might simply be due to the increase in neighborhood information, but could also be a random effect. However, the point where a neighborhood increase might become detrimental has apparently not yet been reached.

As runtime efficiency is also affected by n_{neigh}, runtimes were calculated for 1,000

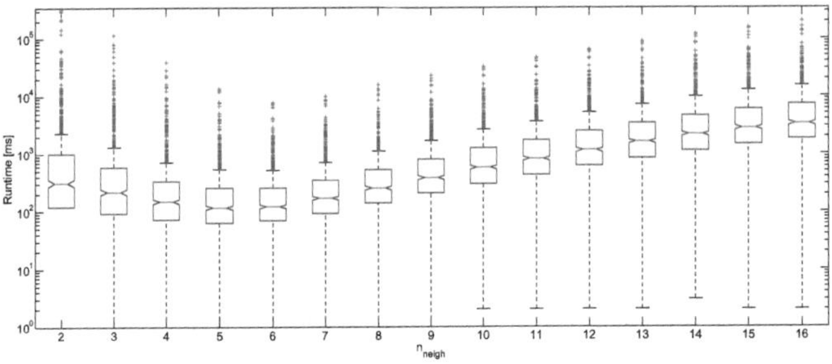

Figure 5.7: Runtimes obtained for 1000 random comparisons for different values of n_{neigh}

	n_{neigh}						
	2	3	4	5	6	7	8
μ	3.6269	1.4160	0.5749	0.3172	0.2814	0.3656	0.5529
σ	2.2753	0.6996	0.2196	0.0925	0.0650	0.0776	0.1139

	n_{neigh}							
	9	10	11	12	13	14	15	16
μ	0.8179	1.2414	1.7499	2.4743	3.3990	4.5653	5.8799	7.0440
σ	0.1653	0.2460	0.3484	0.4900	0.6717	0.9020	1.1534	1.3680

Table 5.4: Runtime requirements [s] of SEGA on the test dataset for different values of n_{neigh}.

comparisons of randomly drawn cavities. Fig. 5.7 shows Box-Whisker plots for elapsed calculation times obtained with different n_{neigh}, the mean runtime is shown in Table 5.4. Not surprisingly, runtime increases with n_{neigh}, favoring a low setting in order to maximize runtime efficiency. Perhaps more unexpected is the observation that runtime requirements at first decrease until $n_{neigh} = 5$. The most likely explanation is that for very low n_{neigh}, the neighborhood size is too small to derive a meaningful local similarity measure. As a result, many ambiguities arise, which have to be resolved by drawing upon global information, which is more expensive than simply realizing unambiguous assignments.

Since the difference in accuracy is not large for $n_{neigh} > 8$ but runtime increases with n_{neigh}, n_{neigh} will be set to 10 for the remaining experiments, as a good compromise between runtime efficiency and accuracy. The same setting was used for SEGAHA to ensure that both variants use the same information.

5.2.2 Influence of the scoring parameter

As mentioned earlier, it is unclear which measure is more suitable for the task of comparing protein binding sites. To get an idea of the influence of α on the predictive performance of both the global as well as the semi-global approaches, the predictive performance of both algorithms were examined with respect to different α.

To this end, pairwise classification were carried out on the four-class dataset used in the previous experiment, this time with a fixed parameter setting but variable α. Each pair of classes of this dataset was analyzed separately, comparing each class with every other. This was done to assess whether one specific setting of α proves superior on different problems.

Intuitively, one would assume that a conjunctive measure would perform better for groups of globally similar binding sites with a high degree of structural conservation, whereas the disjunctive measure should be more suitable if only a fraction of the binding sites correspond to each other. In the latter case, the non-matching part of the larger cavity will have less influence on the similarity score.

In this respect, solving all two-class problems separately might be particularly interesting, since three classes consist of binding sites that interact with rather flexible ligands (ATP, NAD, FAD), whereas the fourth group consists of binding sites with rigid porphyrine rings. Hence, it is more likely for the first three classes to show structural diversity, since the ligands can be bound in different conformations. The fourth class, on the other hand, should be more homogeneous, as the porphyrines will not vary in conformation and thus are more likely to enforce a certain topology of the binding site. Of course, whether this is actually the case strongly depends on the directional interactions between the ligands and the cavities.

5.2.2.1 Influence of the scoring parameter on the performance of SEGA

Classification experiments were again carried out for each two-class problem using SEGA with $n_{neigh} = 10$. Results were obtained by 10-fold stratified cross validation using a k-nearest neighbor classifier and compared with respect to different α. Performance was assessed in terms of classification accuracy. Fig. 5.8 shows the mean accuracy for different values of α, depicted are the results for $k = 1$.

In this case, no single α value is preferable. In some cases, α apparently has no great influence on the classification performance, while in the case of the ATP/FAD and the NADH/FAD problem, a high α appears to be detrimental, thus favoring one-sided

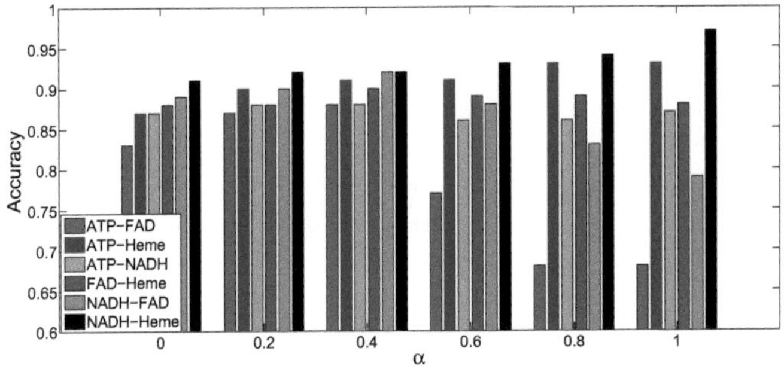

Figure 5.8: Mean MCR on different two-class problems for different α derived from 10-fold stratified cross validation using SEGA ($n_{neigh} = 10, k = 1$).

inclusion. This might indicate the proteins of these classes are more heterogeneous in structure, which is not unreasonable, given that FAD, ATP and NADH binding proteins carry out a variety of different functions and exhibit different folds.

In case of pairwise classification involving heme, a tendency towards mutual inclusion can be observed, although the effect is minimal. This might be explained by the porphyrine-containing pockets being more homogeneous in size and structure, thus a mutual inclusion is more likely to occur in this class. Apparently, α is a problem-specific parameter.

5.2.2.2 Influence of the scoring parameter on the performance of GAVEO

Similar experiments were carried out for the GAVEO and GAVEOc approach, since the corresponding similarity measure (4.9) is again a conjunction of two extremes controlled by a trade-off parameter α. The algorithms were parametrized according to the parameter setting obtained previously. Classification results were obtained from 10-fold stratified cross validations using k-nearest neighbor classification. The results are given in Fig. 5.9 and Fig. 5.10.

As expected, both GAVEO variants yield similar results, in both cases indicating a slight preference for low α values. In general, however, the influence of α appears to be low. Note that the measure used here is a similarity measure instead of a distance measure, as was the case in the previous section. Thus, a low value of α here indicates a preference for the conjunctive measure.

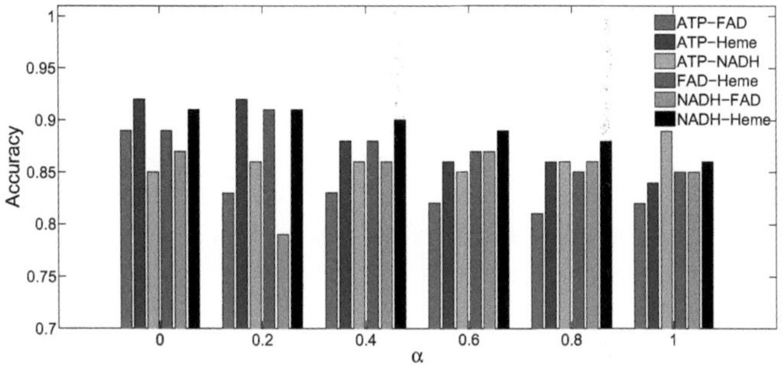

Figure 5.9: Mean MCR on different two-class problems for different α derived from 10-fold stratified cross validation.

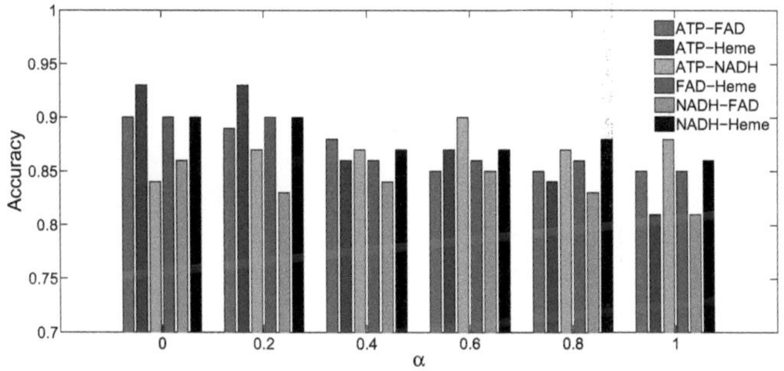

Figure 5.10: Mean MCR on different two-class problems for different α derived from 10-fold stratified cross validation.

Apparently, in case of the GAVEO approach, the performance does not vary largely for different two-class problems, as opposed to SEGA. While this seems to be at odds with the previous results, one has to be aware of the fact that GAVEO measures a global alignment of graphs, i.e., the similarity of the graphs as a whole, whereas the SEGA measure quantifies the aggregated distances between local subgraphs. Thus, the results here are not really comparable to the results of the previous section, since both measures are based on two different concepts of similarity.

5.3 Statistical significance

A typical application of the methods presented here is the retrieval of similar binding sites from a database, given a query structure as reference. Similarity scores obtained by comparing the query to all structures in the database can be used to construct a ranking of similar binding sites, starting with the most similar structures. In such retrieval settings, one is typically interested only in the top x ranks, e.g., in the case of web searches, where one usually focuses on the first retrieved pages.

However, if one is interested in recovering unexpected similarities, e.g., in order to predict cross reactivities for a certain drug, these top results are arguably less interesting than lower ranks. The first ranks will most likely contain structures derived from the same target, but an unrelated protein that is nevertheless similar enough to be addressed by the same ligand as the query might occupy a lower rank with a smaller similarity score.

As mentioned in Chapter 3.4, any score used to compare proteins or protein substructures must be judged against the likelihood that a given score arises by chance, hence a measure of significance is necessary to interpret the results from such a ranking. Given such a confidence measure, one can decide up to which rank the obtained similarity score is still significant enough to be considered, allowing for a more thorough exploration of the retrieved results.

To obtain a significance measure for the similarity scores produced by the different methods, an empirical approach was chosen. For each method, 10,000 randomly drawn comparisons were calculated and the obtained scores were used to derive a score distribution, to which a generalized extreme value distribution (GEVD) was fitted.

As outlined in Chapter 3, one can distinguish three different types of extreme value distributions (EVD): The Gumbel family (type I), the (Frechét) family (type II) and the Weibull family (type III). Since the type of the EVD that should ideally be fitted to the score distributions obtained the various methods was not known in advance, the scores were used to estimate the parameters of a GEVD by maximum likelihood estimation. Subsequently, the corresponding cumulative distribution was used to calculate p-values for the comparison scores.

5.3.1 Statistical significance of the GAVEO score

Fitting a generalized extreme value distribution requires the estimation of three parameters: the shape parameter ξ, the location parameter μ and the scale parameter σ. Thus, 10,000 comparisons should provide a sufficiently large base for the estimation.

	ξ	σ	μ
estimate	-0.220	0.295	0.515
confidence interval	[-0.225, -0.216]	[0.291, 0.300]	[0.509, 0.521]

(a) GAVEO

	ξ	σ	μ
estimate	-0.303	0.298	0.520
confidence interval	[-0.310, -0.296]	[0.293, 0.302]	[0.513, 0.526]

(b) GAVEOc

Table 5.5: GEVD parameter estimates for the score distributions of 10,000 random comparisons using GAVEO and GAVEOc ($\alpha = 0$).

Table 5.5 shows parameter estimates for GAVEO and GAVEOc obtained by fitting a GEVD to the score distribution, along with the corresponding confidence intervals. A parameter lies within the corresponding confidence interval with a probability of 0.95. The probability density function corresponding to the fit of GAVEOc is shown in Fig.5.11a. For reason of comparison, a histogram of the score distribution is shown as well. To convey an impression of the goodness-of-fit of the estimated EDV, the empirical cumulative distribution (green) and the estimated cumulative distribution (blue), from which p-values will be obtained, is also shown in Fig. 5.11b. Similar results were obtained for GAVEO scores, but were omitted here for the sake of brevity.

Since the obtained score distribution obviously depends on the parameter α, EVDs were fitted individually for different values of α and the p-values obtained from the estimated cumulative density functions will be used as measure of significance where appropriate. Table 5.5 shows the estimates for $\alpha = 0$, the complete set of parameter estimates for all α can be found in Appendix B, Table B.6 and Table B.7.

In each case, a type III (Weibull) distribution was obtained, regardless of the choice of α for both GAVEO and GAVEOc. The small confidence intervals and the high similarity between empirical and estimated distribution indicate a reasonably good model of the real score distribution. This is supported by Fig. 5.11b, which shows a nearly perfect fit between the empirical and the estimated cumulative density function. As can be seen in Fig.5.11a, the score distribution is well approximated by the estimated probability distribution. For the following experiments, p-values calculated by the obtained cumulative density functions for the different α settings were used as a measure of significance for the obtained similarity scores where needed.

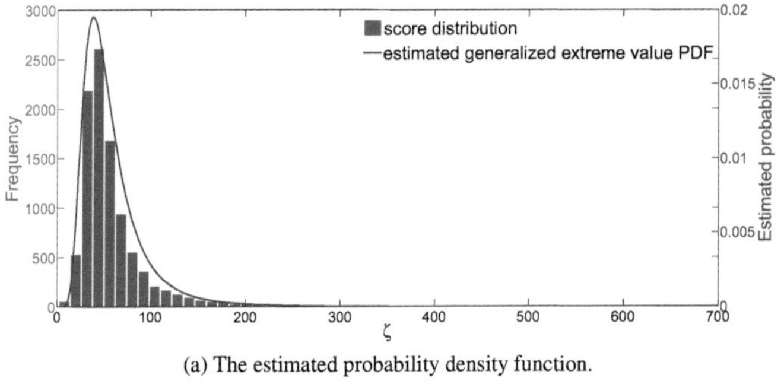

(a) The estimated probability density function.

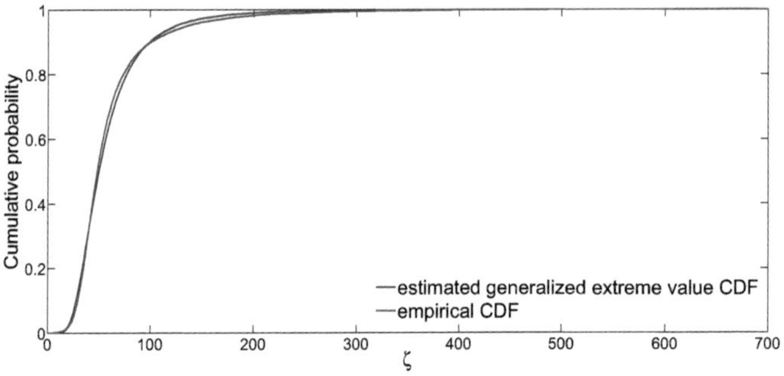

(b) The estimated and empirical cumulative density functions.

Figure 5.11: Visualization of the estimated GEVD for the GAVEOc approach ($\alpha = 0$).

5.3.2 Statistical significance of the local approaches

Similarly, GEVD parameter were estimated for the local approaches, again based on 10,000 comparisons of randomly drawn cavities. The estimated probability density function for BFPJ is shown along with the underlying score distribution in Fig. 5.12a. Again, the empirical cumulative density function is plotted together with the estimated CDF in Fig 5.12b. For the sake of brevity, results are only shown for BFPJ, however, fitting the GEVD for the other fingerprint approaches yielded similarly good estimates.

The obtained parameter estimates for the BFPJ fingerprints are given in Table 5.6, estimates for the other fingerprint approaches can be found in the Table B.5.

As can be seen, the obtained models approximate the underlying score distribution

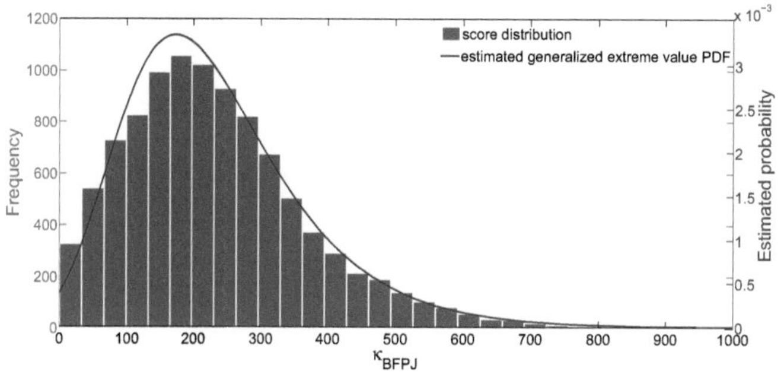

(a) The estimated probability density function.

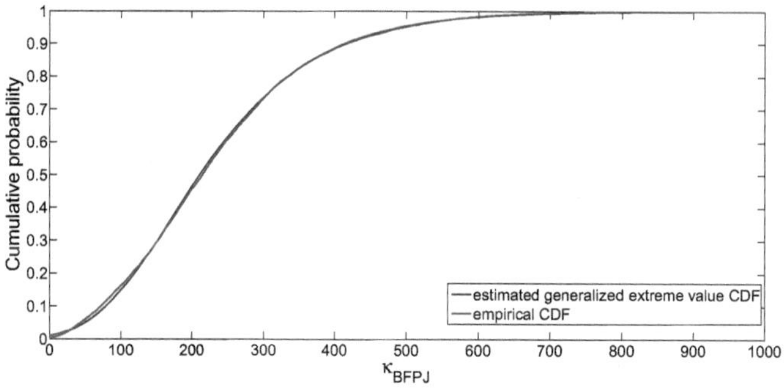

(b) The estimated and empirical cumulative density functions.

Figure 5.12: Visualization of the estimated GEVD for the BFPJ approach.

	ξ	σ	μ
estimate	-0.018	109.838	171.267
confidence interval	[-0.033, -0.003]	[108.074, 111.631]	[168.826, 173.708]

Table 5.6: GEVD parameter estimates for the score distributions of 10,000 random comparisons using the BFPJ fingerprint approach.

	ξ	σ	μ
estimate	0.000	0.784	1.841
confidence interval	[-0.032,0.032]	[0.756,0.812]	[1.802,1.879]

(a) SEGA

	ξ	σ	μ
estimate	-0.024	0.832	1.289
confidence interval	[-0.035,-0.012]	[0.819,0.845]	[1.271,1.307]

(b) SEGAHA

Table 5.7: GEVD parameter estimates for the score distributions of 10,000 random comparisons using SEGA and SEGAHA($\alpha = 0$).

reasonably well, as indicated by the small confidence intervals and the graphical evaluation. For BFPJ, BFPH and FPH, estimating the GEVD yielded type III (Weibull) distributions while for FPJ and FFP type II (Frechét) EVDs were fitted. For the following experiments, p-values calculated by the obtained cumulative density functions were used as a measure of significance for the obtained similarity scores where needed.

5.3.3 Statistical significance of the SEGA score

Finally, GEDVs were estimated for the semi-global approaches SEGA and SEGAHA. Since also the SEGA and SEGAHA scores will depend on the choice of α, the size, scale and location parameters were fitted individually for different values of α. Table 5.7 shows the results of the parameter estimation for $\alpha = 1$, the complete set of estimates for other α can be found in Table B.9 and Table B.8, Appendix B.

Regardless of the choice of α, each estimate yielded a type III (Weibull) EVD for the SEGA approach, with the exception of $\alpha = 0$, where a type I (Gumbel) EVD was obtained. As an example, Fig. 5.13a shows a histogram of the scores obtained for $\alpha = 1$, together with the estimated type II EVD (blue). The empirical (blue) and the estimated (green) cumulative density function is plotted in Fig. 5.13b.

In case of the SEGAHA approach, the obtained EVD were of the Weibull family, except for lower values of α ($\alpha = 0, 0.1$), where type II (Frechét) EVD were obtained.

Again, the low confidence intervals indicate a nearly perfect fit, which can be observed in Fig. 5.13. For the following experiments, p-values calculated by the obtained cumulative density functions (based on the choice of α) were used as a measure of significance where necessary.

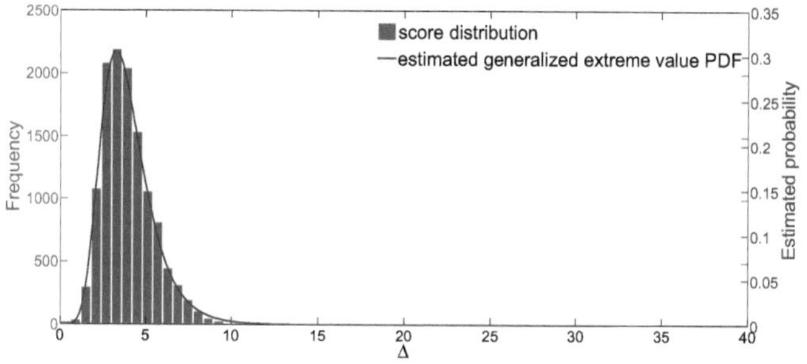

(a) The estimated probability density function.

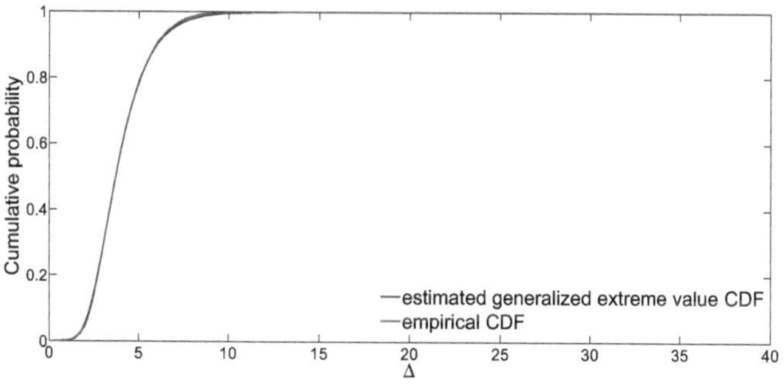

(b) The estimated and empirical cumulative density functions.

Figure 5.13: Visualization of the estimated GEVD for the SEGA approach ($\alpha = 0$).

5.4 Runtime comparison

Since protein structure comparison can potentially be applied in the context of database retrieval, an important aspect of algorithmic performance is the runtime behavior of the different approaches. While the complexity of the different approaches has already been discussed in the method section, an empirical comparison of runtime behavior remains to be done.

To this end, the runtimes were calculated for the different approaches based on a set of 1,000 randomly drawn pairwise comparisons[3]. The resulting times are shown in

[3]Results were obtained on an Intel Core 2 Duo 2.4 GHz, 2 GB memory, Windows XP SP 2 operating system.

	BFPH/BFPJ	BK	FPH/FPJ	FFP	GAVEO	GAVEOc
μ	0.0189	0.0134	0.0205	0.179	5.8410	5.9842
σ	0.0321	0.0497	0.0366	0.663	21.9902	23.3276

	GH	RW	SEGA	SEGAHA	SP	SPSA
μ	0.655	0.2811	0.0185	0.0423	0.0980	5.7440
σ	0.891	0.1288	0.0227	0.0553	0.9780	40.8970

Table 5.8: Mean μ and standard deviation σ for the runtime performance of the different algorithms based on 1,000 comparisons in seconds.

Table 5.8.

In terms of runtime behavior, the crisp fingerprint approaches as well as BK and SEGA show the best performance. GAVEO and GAVEOc have by far the highest runtime, which is not surprising, since evolutionary optimization is known to be expensive. However, the clique-based BK and GH have the drawback of a space complexity of $\mathcal{O}(n^4)$ (n = number of nodes), which is problematic for calculating alignments for large graphs on current machines. The other approaches only have a space complexity of $\mathcal{O}(n^2)$. Moreover, given that the clique-enumeration problem is NP-complete, the good runtime performance of BK is somewhat deceiving, being mainly the result of small cliques and the above-mentioned limitation to the first one hundred cliques.

Among the kernel functions, the shortest path kernel in combination with sequence alignment has a prohibitively high runtime, in magnitude comparable to the evolutionary optimization approaches. Since the main argument for using local methods instead of more complex global algorithms is a potentially efficient runtime behavior, it is obvious that the approach is not useful in the context of protein structure comparison. Hence, it will be excluded from further experiments, with one exception (cf. Section 5.7) serving as a proof-of-principle to compare its performance with the unmodified shortest path kernel.

The other kernels along with the fingerprint approaches are again comparably fast, the random walk kernel being the slowest local approach with an average runtime that is roughly 30 times longer than SEGA. This is not surprising, given its runtime complexity of $\mathcal{O}(n^6)$.

5.5 Tolerance towards structural variation

As outlined in the previous chapter, one motivation for the development of non-global approaches was to derive an algorithm which allows for a greater tolerance towards structural variation when dealing with protein structure data. Of these, only the semi-global SEGA additionally yields a mutual correspondence of pseudocenters. With SEGA, a greater tolerance should be achieved by assembling the alignment from local comparisons, thus theoretically allowing to ignore larger structural variations, as long as the node neighborhood remains largely conserved. However, practically, this remains to be assessed.

The purpose of the following experiment is to compare the performance of the different algorithms with respect to alignment quality when confronted with different types and levels of structural variation. This requires a closer look at the obtained alignments to judge whether the correct correspondences have been recovered. For this reason, the local approaches where excluded from this experiment, as they only produce a degree of similarity rather than an interpretable structural alignment.

5.5.1 Semi-synthetic experiment

To study the effects of variation on the alignment quality in a systematic way, it would be necessary to control the degree of variation distinguishing two binding sites from one another while the correct alignment is still known in advance. As this is hardly possible for real data, a semi-synthetic approach was adopted to compare the robustness of the different approaches towards such inaccuracies. To this end, 100 protein binding pockets co-crystallized with a ligand were randomly selected from CavBase. These pockets have the benefit of being relatively large which is beneficial to generate non-trivial deformations. Based on each binding pocket, ten structurally diverse synthetic pockets where generated by subjecting the original pocket to structural perturbance.

More precisely, two types of variation where considered: mutational variation, affecting pseudocenter labels due to mutations on the amino acid level, and structural variation, affecting the position of pseudocenters. Structural noise was introduced by translating each center by a randomly directed vector with normally distributed length controlled by a deviation parameter $p_{dev} \in [0, 1]$ (standard deviation of the vector length normal distribution). To introduce mutations, all pseudocenters where subjected to label mutation controlled by a mutation parameter $p_{mut} \in [0, 1]$ (mutation probability). The generated binding sites were then aligned to the unaltered template, with the identity alignment (an alignment of each node onto the corresponding node with the same index) assumed as correct.

Pairwise alignments of the binding pockets were subsequently calculated using SEGA, SEGHA and GAVEO[4], as well as BK and GH as baseline approaches. The percentage of correctly matched pseudocenters of the core pocket was determined for different values of p_{dev} and p_{mut}. Fig. 5.14 shows the mean percentage of correctly mapped centers for different algorithms under structural noise (5.14a), mutation (5.14b) and a combination of both (5.14c). Runtime requirements of the approaches are summarized in Table 5.8.

Apparently, SEGA shows the most stable performance, rivaled only by GAVEO, although the runtime requirements of GAVEO are much higher. Yet, when combining both sources of variation, the performance of the evolutionary optimization declines after a certain level of noise, which indicates that the objective function now favors alternative alignments. This is not surprising, since both node labels and edge distances are now affected by alteration. However, an increased number of stall generations might alleviate this effect, assuming that the optimization process has not yet reached a (near-) optimal solution.

The neighbor-based SEGA approach instead is less affected. Interestingly, the quality of the SEGAHA alignments also deteriorates, suggesting that the added stability of SEGA can be attributed to the additional usage of a global structural reference frame for resolving ambiguous assignments, the main difference between the two approaches.

The results indicate that SEGA represents the best compromise between runtime efficiency and robustness, as can be seen in Fig. 5.8. While GAVEO also proves relatively stable when confronted with noise, the runtime requirements for the optimization are much higher. Yet, the more faster approaches are generally also more intolerant towards noise, with the notable exception of the semi-global SEGA.

While the results convey a realistic picture on the tolerance of the different approaches when confronted with increasing levels of noise, one could argue that a semi-synthetic approach might not be relevant for real data, since the level of structural distortion might be much less there. Therefore, another experiment was carried out on a real dataset as well, the Astex non-native dataset. Since this study was carried out as a retrieval experiment, it will be presented in the next section, along with other retrieval experiments.

[4]GAVEOc was omitted, since the evolutionary optimization that allows for tolerance towards structural variation is identical to GAVEO.

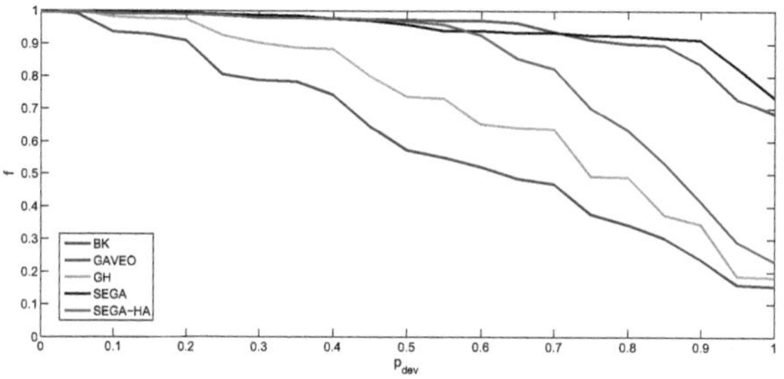

(a) distortion

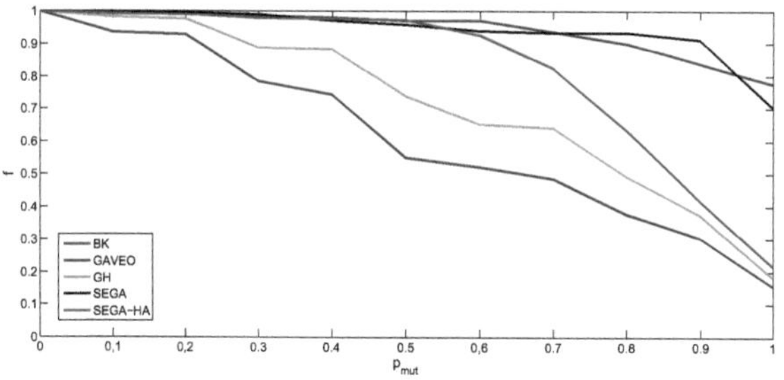

(b) mutation

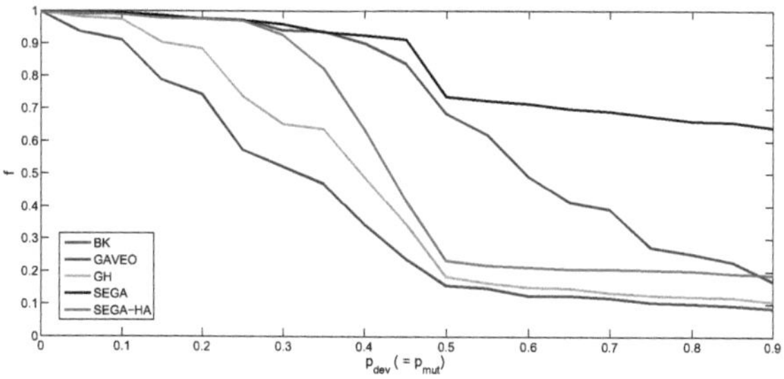

(c) distortion and mutation

Figure 5.14: Relative frequency f of correctly mapped pseudocenters (y-axis) for different types (a) distortion, b) mutation, c) both) and levels of variation

5.6 Similarity retrieval

One major goal of this work was the development of algorithms capable of retrieving similar protein structures or binding sites from a database, given a query of interest. This allows for the identification of functionally related proteins, even if no significant similarity on the sequence level exists, for example when searching for analogous proteins that have developed independently but converged to a similar function. In the context of knowledge-based drug design, this is an important application, since it allows in principle for the identification of potential cross-reactivities, especially when focusing on protein binding sites instead of whole protein structures.

The purpose of the following experiments is to assess the capability of the introduced algorithms to perform this task, i.e., to retrieve similar binding sites in a retrieval scenario. In the first set of experiments, the algorithms were applied on an independent benchmark dataset compiled by the Nussinov-Wolfson group for the evaluation of the SiteEngine approach (Shulman-Peleg et al., 2004). Subsequently, this is taken one step further by using the algorithms in retrieval experiments covering the whole CavBase.

5.6.1 Similarity retrieval on a benchmark dataset

Conducting the first set of retrieval experiments on the predefined SiteEngine benchmark set offers three benefits. Firstly, as was shown in the preliminary experiments, some of the algorithms have relatively high runtime requirements. Conducting initial retrieval experiments on a smaller benchmark set allows one to gain first impressions on the retrieval performance of the algorithms without excessive runtime investment.

Secondly, assessing the performance of the introduced algorithms in a retrieval experiment is not as trivial a task as it appears. Given that each of the algorithms allow for some degree of structural variation, some more than others, the question arises how the results should be interpreted. For example, SEGA and SEGAHA were developed to allow for a greater structural tolerance than provided by a simple clique search, allowing for the detection of more remote similarities. With that idea in mind, it becomes obvious that classical methods to assess structural similarity, such as the RMSD value, would be too stringent for this purpose. Also, simply looking for the same target is too restrictive for the purpose of identifying potential cross reactivities. However, for the benchmark set, the classes were already defined in a previous study (Shulman-Peleg et al., 2004).

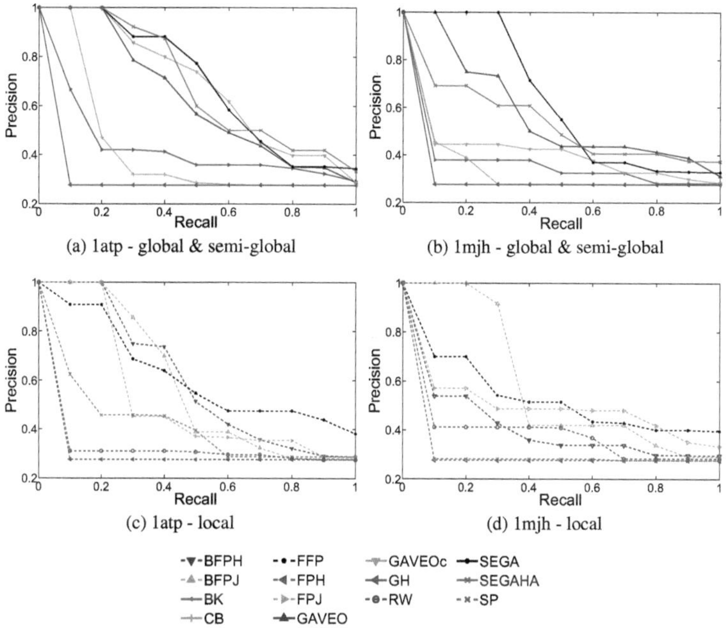

Figure 5.15: 11-point precision-recall curves on the SiteEngine dataset for the query proteins 1atp and 1mjh.

A third advantage is the fact that the benchmark dataset allows for a more in-depth analysis of the results, as the dataset is small enough to allow a manual inspection and interpretation of the results. As query structures, the same proteins were selected that were also used in the original SiteEngine retrieval experiments to calculate similarity rankings: the adipocyte lipid-binding protein from *M. musculus* (PDB code: 1lib), a CAMP-dependent protein kinase from *M. Musculus* (PDB code: 1atp), the hypothetical protein MJ0577 from the genome of *M. jannaschii*, which contains an ATP binding site (PDB code: 1mjh) and the human sex hormone-binding globulin (SHBG) complexed with the steroid estradiol (PDB code: 1lhu). Additionally, two other proteins from the group of estradiol binding proteins were selected as queries: the human estrogen receptor (PDB code: 1ere) and a mutant human nuclear estrogen receptor (PDB code: 1qkt).

Retrieval experiments were carried out by comparing the main pocket of each protein

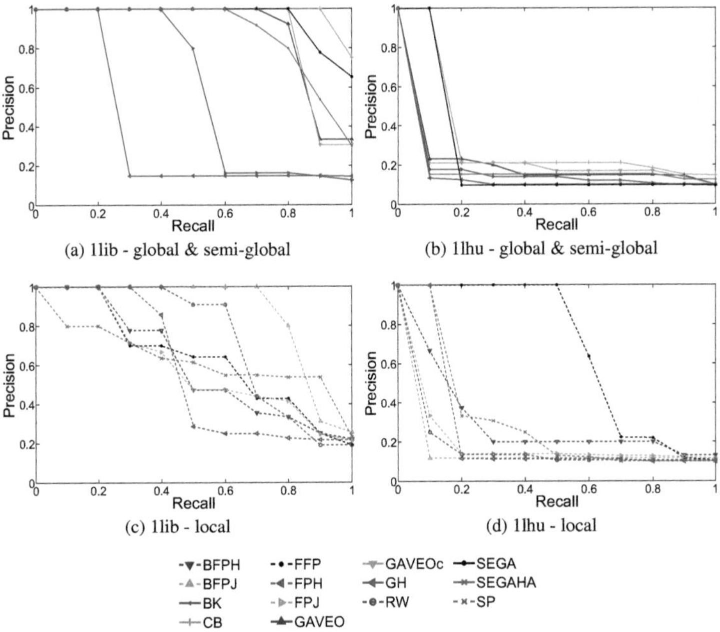

Figure 5.16: 11-point precision-recall curves on the SiteEngine dataset for the query protein 1lib and 1lhu.

to each query binding site, deriving a ranking based on the obtained similarity scores. For each ranking, 11-point interpolated precision-recall curves[5], commonly used for the assessment of retrieval systems (Manning et al., 2008), were derived, depicted in Fig. 5.15 to Fig. 5.17.

Among the local approaches, BFPJ and FFP achieve the best results, with BFPJ showing superior performance on four of the six queries while FFP yields a better ranking for two of the queries: 1lhu and 1ere. Interestingly, in these two cases, none of the other local approaches is able to achieve reasonably good results, suggesting that the problem of discontinuity might indeed have a strong effect in some cases.

[5] 11-point precision-recall curves depict the observed precision at fixed recall levels in steps of 0.1 and interpolate the intervals in between.

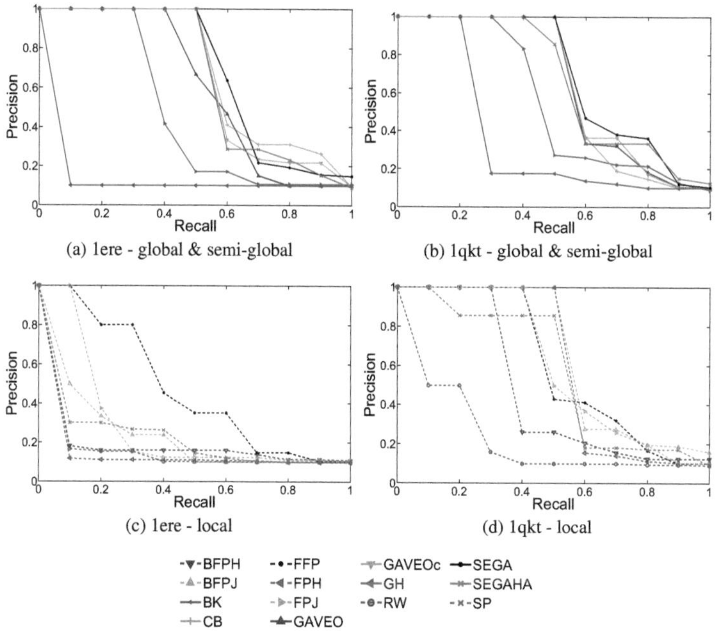

(a) 1ere - global & semi-global (b) 1qkt - global & semi-global

(c) 1ere - local (d) 1qkt - local

Figure 5.17: 11-point precision-recall curves on the SiteEngine dataset for the query protein 1ere.

Especially for 1lhu, the other approaches, global and local, seem to be unable to achieve a good result, suggesting that there might be no recognizable structural resemblance between the query structure and other estradiol binding pockets in the dataset. This was not the case in the original SiteEngine experiment (Shulman-Peleg et al., 2004), which could be attributed to the differences in pocket extraction and representation between CavBase and SiteEngine. For this reason, two other estradiol binding query structures (1ere, 1qkt) were included, to assess whether a meaningful ranking can be obtained within this class at all. Apparently, this is possible, since the results for the other queries are quite good.

In most cases, the Jaccard measure tends to yield better results than the Hamming distance, regardless of the fingerprint model, similar to the observation made on the four-class classification dataset in Section 5.2.1.2.

Rank	PDB	Function	Score	P-value
1	1lib	Adipocyte lipid-BP	90.00	$< 10^{-10}$
2	1lie	Adipocyte lipid-BP	41.04	$< 10^{-10}$
3	1lid	Adipocyte lipid-BP	40.16	$< 10^{-10}$
4	1hms	Heart muscle fatty acid-BP	20.72	$3.96 \cdot 10^{-10}$
5	1b56	Epidermal fatty acid-BP	10.14	$2.79 \cdot 10^{-5}$
6	1pmp	Myelin P2	10.09	$2.94 \cdot 10^{-5}$
7	1ftp	Locus muscle fatty acid-BP	9.13	$8.88 \cdot 10^{-5}$
8	1opb	Cellular retinol BP II	8.61	$21.61 \cdot 10^{-4}$
9	1cbs	Cellular retinoic acid BP II	8.19	$2.61 \cdot 10^{-4}$
10	2cbr	Cellular retinoic acid BP II	7.30	$7.17 \cdot 10^{-4}$
11	1opa	Cellular retinol BP II	7.00	$1.06 \cdot 10^{-3}$
12	1kqw	Cellular retinol BP	6.60	$1.70 \cdot 10^{-3}$
13	1mdc	Insect fatty acid BP	5.97	$3.65 \cdot 10^{-3}$

Table 5.9: Ranking of comparisons between the fatty acid binding protein 1lib and the SiteEngine benchmark set ($\alpha = 1$).

Among the global and semi-global approaches, the SEGA algorithm generally shows the best performance, with one notable exception for the query structure 1lib. While SEGA's performance is still strikingly good, the CB approach is capable of retrieving an even better result, although it fails to outperform SEGA on the other queries. However, the difference is not large: SEGA was able to retrieve 13 of the 15 fatty acid-binding proteins present in the dataset exclusively on the top ranks with significant p-values, while CB retrieved one more structure.

Nevertheless, it is interesting to observe that both algorithms, as well as GAVEO are able to retrieve such good results, since some of these structures were not retrieved in the original SiteEngine experiment (Shulman-Peleg et al., 2004). SiteEngine retrieved proteins from other classes among the top ten results (ketosteroid isomerase, HIV protease and others), which are not present in the top ranks here and achieve non-significant scores for SEGA and GAVEO. Yet, this is not a fair comparison, since SiteEngine is based on a slightly different modeling concept, which is why this approach was not used as a baseline here. As an example, the ranking obtained by SEGA is shown in Table 5.9.

While SEGA generally shows the best performance, GAVEO is not much worse in most cases, although the results are obtained with a much higher runtime expenditure. It is also notable, that both global approaches (GAVEO and GAVEOc) as well as both semi-global approaches (SEGA and SEGAHA) are able to improve upon the baseline algorithms. The only real competitor approach appears to be the original CavBase approach

(CB), which nevertheless fails to surpass the best global and semi-global algorithms with the one exception mentioned above.

The results indicate that on average, the global and semi-global methodologies retrieve more relevant structures than the local methods, although exceptions exist, e.g., for the query 1mjh, where GAVEO does not retrieve relevant structures on the first ranks. For this query, GAVEOc shows a much better performance, although in general GAVEO is more successful. This suggests that preserving the clique solution during the optimization process can sometimes lead to better results although the majority of cases indicate the opposite.

5.6.2 Tolerance toward structural variation and retrieval performance on real data

The semi-synthetic experiment outlined above has demonstrated the robustness of SEGA and, to a lesser extend, GAVEO towards structural noise and mutational variation. However, one could argue that this increased tolerance might not be relevant when working on real data, as the structural variation present in related binding sites might be far less pronounced.

Moreover, so far, it is not clear how well the local approaches will perform when confronted with structural variation. On the one hand, these approaches should be much less affected by structural changes, on the other hand, as outlined in Chapter 4, local approaches might suffer from a large false positive rate, which would again lead to deteriorated results.

To shed some light on these issues, the Astex non-native dataset was used in addition, which contains contains roughly 1,000 structures in total representing different conformations of the 65 different protein targets (apo structures and complexes with other ligands). Thus, this dataset presents a representative selection of typical variation observed in real structures, albeit for identical targets.

Again, retrieval experiments were carried out, with each of the 65 target structures as queries. More precisely, the main pockets containing the co-crystallized ligands were retrieved from CavBase and used as queries against the complete set of cavities belonging to the non-native structures. The similarity of two protein structures is given by the similarity between the query cavity and the most similar cavity with a different conformation.

This is a valid strategy also common in other bioinformatic applications, e.g., local sequence alignment, where the similarity score corresponds to the similarity between the

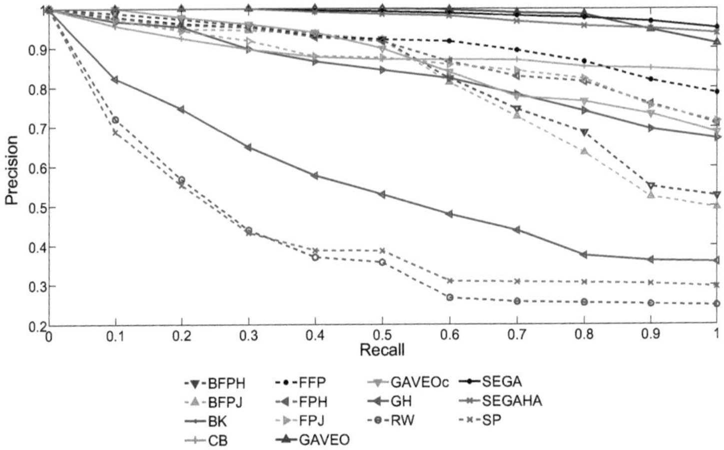

Figure 5.18: Averaged 11-point precision/recall curves for the different algorithms and competitor approaches on the Astex non-native dataset.

two most similar subsequences. Thus, there is no need for preselecting the true protein binding site for all query proteins. While this might still be possible on the Astex dataset, it would not be feasible for a retrieval on the complete CavBase.

For each of the approaches, rankings were calculated based on the associated similarity/distance measure and subsequently filtered to contain only one cavity per structure (keeping the most similar ones). The derived rankings for all 65 queries were then used to calculate average 11-point precision-recall curves, averaging over all queries, a procedure commonly used for the comparison of retrieval systems over multiple queries (Bustos et al., 2004; Manning et al., 2008). The results are shown in Fig. 5.18.

The results nicely demonstrate that dealing with structural differences is not a trivial problem, even for the same target protein, since none of the approaches is able to retrieve all similar structures in all cases. The highest performance is achieved by GAVEO, SEGA and SEGAHA, which are able to outperform all competitor approaches as well as the local methods. SEGA arguably shows the best performance, as all relevant structures are placed among the top ranks, indicated by the slow declining tail of the curve. On the other hand, the GAVEO approach seems to be slightly more successful at medium recall levels, indicating that SEGA retrieves slightly more false positives here. However, the difference is minimal.

The fingerprint approaches along with GAVEOc and the best competitors CB and BK show medium performance, with a much more stable performance than the remaining competitors (GH, SP and RW). Again, it can be observed that preserving the clique solution during evolutionary optimization by GAVEOc does not have a beneficial effect and can even be detrimental. Still, GAVEOc as well as the fingerprint methods are capable of outperforming all competitors, even the original CavBase algorithm (CB) for low to medium recall levels. Unfortunately, the curves decline more strongly at high recall levels, indicating that some relevant structures are ranked relatively low. For comparison, CB places the last relevant structure on a much higher rank.

Unfortunately, the fingerprint approaches fail to attain similarly strong results as the global GAVEO and the semi-global SEGA and SEGAHA, the stronger decline of precision indicating a larger number of false positives, confirming the apprehensions raised earlier. Among the local approaches, the best performance for low to medium recall levels is achieved by BFPH and BFPJ, which is in accordance with the results obtained earlier. However, for higher recall levels, performance deteriorates more strongly than for the ε-based fingerprints (FPH and FPJ). The most stable of the fingerprint approaches is the fuzzy fingerprint approach.

To condense the comparison of the approaches to a few single values, four different performance measure were calculated, commonly encountered in the field of retrieval systems. The $P10$ measure is given by the precision of the approaches after ten retrieved items (Manning et al., 2008). This measure is typically used under the premise that users of retrieval systems expect to find the relevant item among the top ten results, and thus correlates well with user satisfaction. While it is easy to interpret, it averages poorly over different queries due to the fixed rank and has a larger margin of error.

As alternative measure, the R measure (Manning et al., 2008) and the MAP measure (Manning et al., 2008) were also used, both having a lower margin of error than $P10$ (Buckley and Voorhees, 2000). The R measure corresponds to precision after retrieved r items, r being the number of relevant items in the dataset. MAP (Mean Average Precision) is given by the average of precision values after each relevant item is retrieved. MAP uses more information than the $P10$ or R and is more stable with an even lower margin of error than R (Buckley and Voorhees, 2000), yet it is not easily interpreted, since a low MAP score can arise from several reasons.

As a fourth measure, B_{Pref} is used, which is more stable for cases where the relevance judgment is incomplete or erroneous, i.e., cases where relevant items are judged as irrelevant (Buckley and Voorhees, 2004). This is an important advantage for judging the

Algorithm	$P10$	R	MAP	B_{Pref}
BFPH	0.56	0.60	0.62	0.60
BFPJ	0.55	0.60	0.61	0.60
BK	0.56	0.65	0.66	0.65
CB	0.56	0.70	0.71	0.70
FFP	**0.60**	**0.74**	**0.75**	**0.74**
FPH	0.58	0.69	0.70	0.69
FPJ	0.56	0.68	0.69	0.68
GAVEO	**0.65**	**0.81**	**0.81**	**0.81**
GAVEOc	0.59	0.67	0.68	0.67
GH	0.30	0.34	0.36	0.34
RW	0.15	0.19	0.20	0.18
SEGA	**0.63**	**0.80**	**0.79**	**0.79**
SEGAHA	**0.63**	**0.80**	**0.79**	**0.79**
SP	0.13	0.21	0.23	0.20

Table 5.10: Performance of the different approaches in terms of $P10$, R, MAP and B_{Pref}. Good performances are in bold face.

performance of similarity retrieval of protein binding sites, since, as was already stated above, unexpected similarities might arise, which are not easily interpreted. This is more relevant in the next section, where retrieval experiment on the complete CavBase were performed. The four evaluation criteria for the Astex retrieval experiments are given in Table 5.10.

The results confirm the impression conveyed by Fig. 5.18, all four measures showing a similar picture. The best results are achieved by GAVEO, SEGA, SEGAHA and, to a lesser extend the fuzzy fingerprint approach, while the greedy heuristic, as well as the random walk and shortest path kernel perform extremely weak.

It can also be observed that the $P10$ measure, as argued above, is less discriminant, showing for example no difference between *CB* and the other fingerprint approaches, while the other, more stable measures show a better performance of the competitor approach of CavBase, which is in good agreement with the average precision-recall curves.

Despite the more stable precision-recall curve of SEGA, GAVEO yields slightly higher performance values, reflecting the better performance for medium recall levels, since this difference is also apparent for the $P10$ measure. This is at odds with the synthetic experiment above suggesting a slightly higher tolerance of the SEGA approach towards structural differences compared to GAVEO. Thus, a closer look at the results was taken to find the reason for this slight discrepancy.

Rank	PDB	Function	Score	P-value
1	1oiu	CDK2 *H. sapiens*	15.63	$3.8 \cdot 10^{-8}$
2	1v0b	PfPK *P.falciparum*	14.68	$1.3 \cdot 10^{-7}$
3	1oi9	CDK2 *H. sapiens*	13.60	$4.8 \cdot 10^{-7}$
4	1h1r	CDK2 *H.sapiens*	13.24	$7.6 \cdot 10^{-7}$

(a) 1v0p

Rank	PDB	Function	Score	P-value
1	1ke6	CDK2 *H. sapiens*	31.56	$1.1 \cdot 10^{-16}$
49	1ob3	PfPK *P.falciparum*	6.82	$1.6 \cdot 10^{-10}$
51	1v0o	PfPK *P.falciparum*	6.52	$2.4 \cdot 10^{-10}$
58	1v0b	PfPK *P.falciparum*	5.68	$7.4 \cdot 10^{-10}$

(b) 1ke5

Rank	PDB	Function	Score	P-value
1	1mvt	DHFR (*H. sapiens*)	45.73	$< 10^{-16}$
2	1mvs	DHFR (*H. sapiens*)	24.12	$1.2 \cdot 10^{-12}$
3	1aoe	DHFR (*C. albicans*)	19.28	$4.2 \cdot 10^{-10}$
5	1ia2	DHFR (*C. albicans*)	18.50	$1.2 \cdot 10^{-9}$

(c) 1s3v

Rank	PDB	Function	Score	P-value
1	1m78	DHFR (*C. albicans*)	67.49	$< 10^{-16}$
1	1m78	DHFR (*C. albicans*)	62.50	$< 10^{-16}$
6	1mvt	DHFR (*H. sapiens*)	14.17	$7.5 \cdot 10^{-7}$
7	1mvs	DHFR (*H. sapiens*)	14.05	$8.7 \cdot 10^{-7}$

(d) 1ia1

Table 5.11: Examples of retrieved proteins for the queries 1v0p, 1ke5, 1ia1 and 1s3v using the SEGA approach.

Upon inspection, SEGA shows decreased precision levels in only four cases, or, more appropriate two pairs: The two query structures 1v0p and 1ke5 as well as 1s3v and 1ia1. In each case, members of the former class are ranked high in the latter and vice versa, with significant scores, as shown in Table 5.11. This oddity raises the question whether there is some actually resemblance between these targets or whether these are simply false positives. Indeed, 1v0p is a crystal structure of the *P. falciparum* protein kinase 5 (PfPK5), which is complexed with purvalanol B, while 1ke5 represents the human cyclin-dependent kinase 2(CDK2). Purvalanol B, however is a potent inhibitor of protein kinases, which is known to inhibit both PfPK5 and human CDK2 (Villerbu et al., 2002).

Algorithm	$P10$	R	MAP	B_{Pref}
BFPH	0.567	0.598	0.612	0.597
BFPJ	0.552	0.601	0.608	0.597
BK	0.570	0.651	0.664	0.649
CB	0.567	0.703	0.714	0.707
FFP	0.611	0.738	0.749	0.737
FPH	0.583	0.693	0.701	0.690
FPJ	0.564	0.682	0.694	0.679
GAVEO	**0.659**	**0.814**	**0.821**	**0.816**
GAVEOc	0.602	0.664	0.670	0.663
GH	0.288	0.329	0.313	0.303
RW	0.148	0.193	0.201	0.180
SEGA	**0.661**	**0.824**	**0.834**	**0.830**
SEGAHA	**0.658**	**0.814**	**0.825**	**0.819**
SP	0.125	0.212	0.230	0.202

Table 5.12: Corrected performance of the different approaches in terms of $P10$, R, MAP and B_{Pref}.

In case of the second pair, 1s3v corresponds to human dihydrofolate reductase (DHFR) while 1ia1 is a crystal structure of DHFR from *Candida albicans*. Both queries are complexed with different but chemically related tetrahydroquinazoline antifolates ((2R,6S)-6- [methyl (3,4,5-trimethoxyphenyl) amino] methyl-1,2,5,6,7,8- hexahydroquinazoline-2,4-diamine and 5-phenylsulfanyl -2,4- quinazolinediamine).

Apart from the fact that both queries represent the same enzyme from different species and, as dihydrofolate reductases, use NADPH as a substrate, both enzymes can be inhibited by similar inhibitors. For example, a potent non-selective inhibitor of both targets is 6-substituted-5-(4-tert-butylphenyl)thiol-2,4-diaminoquinazol (Chan et al., 1995). Yet, differences in the binding site geometries also exist and selective inhibitors of DHFR from *C. albicans* are known (Chan et al., 1995).

Given that the ultimate goal of both approaches is the discovery of proteins affected by the same inhibitor or interacting with the same molecules, the results obtained by SEGA cannot be considered unreasonable, since it recognizes the similarities that obviously exist for each of the pairs. Having established this relationship, a more natural interpretation of the results would be to treat these cases as true positives as well, thus correcting the relevance judgment accordingly. A corrected evaluation yields the results shown in Table 5.12.

As one can see, performance values increase for each approach, indicating that the cavities associated with the similar queries indeed do achieve a relatively high rank. After

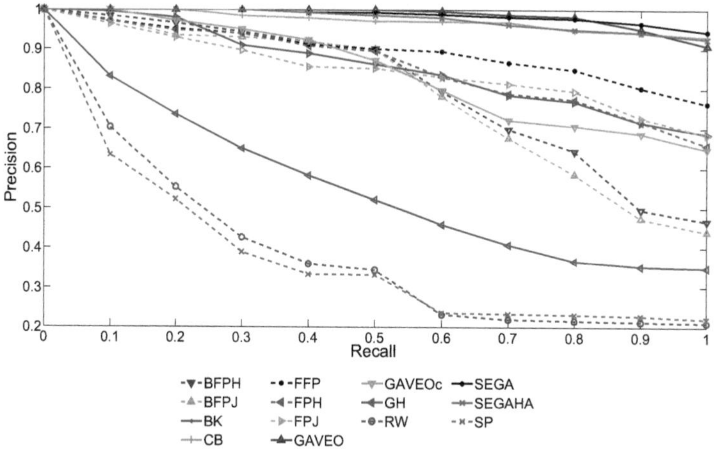

Figure 5.19: Averaged 11-point precision/recall curves for the different algorithms and competitor approaches. The evaluation was limited to those queries, for which CB could calculate all pairwise comparisons.

taking these similarities into account, SEGA yields slightly better results than GAVEO, now performing best, closely followed by GAVEO and SEGAHA.

Upon closer inspection of the results it becomes apparent, that the CavBase approach is unable to calculate pairwise comparisons for several large cavities, due to an exhaustion of runtime memory. This is also true for the RW kernel, which might partly explain its exceedingly weak performance. This might raise the question whether the GAVEO and SEGA approach only show a better performance due to the inability of the CB approach to compare larger binding sites.

Therefore, another set of average precision-recall curves are shown in Fig. 5.19, this time limited to those queries where CB could calculate scores for all pairwise comparisons. In total, 51 out of 65 queries were left for the analysis. The results show that indeed, the performance of CB is better in these cases, being comparable to the SEGAHA approach, though both GAVEO and SEGA still yield better results.

5.6.3 Retrieval of similar binding sites from CavBase

Having compared the approaches on a real benchmark dataset, the methods were employed in a broader context as a next step. This was again realized by retrieval experiments of similar binding sites from CavBase, this time covering the whole "cavity space", i.e., the space of all putative protein binding sites. Since performing retrieval experiments on the whole CavBase will be computationally expensive, at least for some of the approaches, these experiments were limited to the high-resolution subset of CavBase mentioned in Section 5.1, containing all cavities with a minimal resolution of 2.5 Å.

Initially, four exemplary proteins where selected of which the main cavity was used as query: A human carbonic anhydrase II (PDB code: 2eu2), candidapepsin 2 (PDB code: 1eag), an aspartic protase (SAP2) from *C. albicans*, DESC1, a human type II transmembrane serine protease (PDB code: 2oq5) and human MAP kinase 14 (PDB code: 3hec).

The queries were selected to include two large protein groups from the PDB (serine proteases and serine/threonine kinases) as well as members of two smaller groups within the PDB (carbonic anhydrases and aspartic proteases) (Berman et al., 2000).

The group of aspartic proteases can be divided into 16 families or 5 clans according to the MEROPS database (Rawlings et al., 2010). Nearly all aspartic proteases are inhibited by pepstatin (Umezawa et al., 1970), which justifies the assumption, that the active sites of these enzymes share some degree of similarity. Moreover, all aspartic protease share a common binding mode where a protein substrate is bound in a tunnel spanning the whole protein with two aspartates flanking each other in the catalytic center (Klebe, 2009). Thus, this group might be more homogeneous with respect to protein structure.

Serine proteases, on the other hand, are separated into 45 families and 13 clans, the dominant group being chymotrypsin-like peptidases (Rawlings et al., 2010). Serine proteases are known to exhibit different folds but nevertheless share the same function, justifying the assumption, that this group is more heterogeneous. In fact, the classic examples of proteins with different folds carrying out the same function, trypsin and subtilisin, are serine proteases (Schmitt et al., 2002). α-Carbonic anhydrases are largely similar in structure, yet different from other carbonic anhydrases, whereas MAP kinases, or serine/threonine kinases in general exhibit largely the same fold.

Additionally, some of the queries from the SiteEngine dataset were employed again: human adipocyte lipid binding protein (PDB code: 1lib), which was used as an example for fatty-acid binding proteins, and a structure of the human sex-hormon binding protein (PDB code: 1lhu) complexed with estradiol, as an example for a steroid-hormone binding protein. Moreover, thermolysin was selected as query, representing the highly populated group of metalloproteases (PDB code: 1tmn), for which extensive inhibitor studies have been performed, and a structure of the HIV-1 protease (PDB code: 2r38), another aspartic protease representing a prominent drug target for HIV treatment for which a large number of crystal structures are available (Berman et al., 2000).

As mentioned in the previous section, judging the relevance of retrieved results is not a trivial task, given that each of the algorithms allows for some degree of structural variation. As argued above, using RMSD-based evaluation is too restrictive for this purpose. The same holds true for judging only structures of the same target as relevant, since this will obviously never cover cross reactivities.

Another possibility would be to judge the similarity of binding sites according to functional or structural annotation of the corresponding proteins, for example based on the EC nomenclature or the SCOP or CATH classification. While this represents a viable alternative, assessing similarity in this manner means inherently resorting to a different concept of similarity. For example, it is plausible that proteins sharing the same EC class and/or fold could bind identical ligands, while the opposite is not necessarily true. In other words, a structure might belong to a different fold and enzyme class and still be relevant, since it can be affected by the same drug-like molecule.

Thus, with the purpose of this thesis in mind, the obvious alternative would be to consider two structures as similar if they bind the same ligand, as this is the actual property one wants to predict. Unfortunately, this information is not always available for an arbitrary CavBase structure, otherwise there would be no need for the comparison of binding sites in the first place. As each possibility is a viable but incomplete option, results will be evaluated according to the following rule: two proteins will be considered similar if

- the proteins belong to the same SCOP family,
- the proteins are annotated with the same EC number,
- or the proteins are known to bind the same ligand.

The first two conditions are easily verified by consulting the SCOP (Murzin et al., 1995) and the E.C. database (Webb, 1992). The third condition can be verified by consulting the PDB Berman et al. (2000) and PDBbind databases (Wang et al., 2004). The former criteria are still rather conservative, since fold and enzyme class will capture only closely related structures. The third criterion is more useful for judging retrieved proteins with respect to occurring cross reactivities, even for proteins that are not functionally related. However, this obviously requires such information to be available.

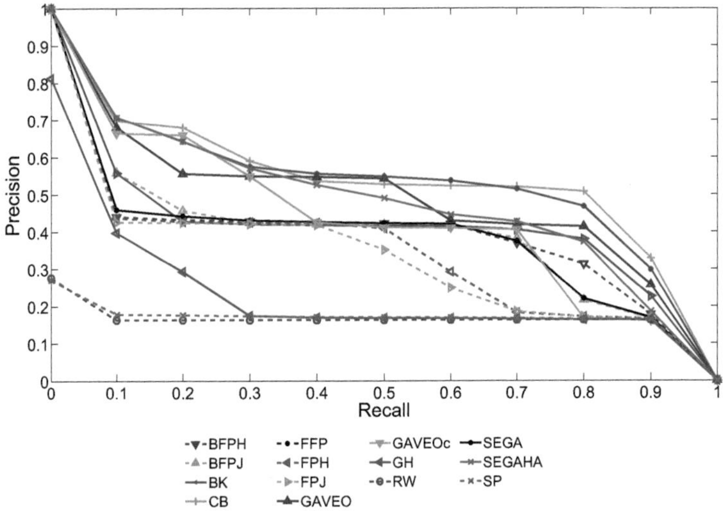

Figure 5.20: Averaged 11-point precision-recall curves for the different approaches based on rankings of all eight query structures.

At this point, it should be stated, that the above criteria are not sufficient to proof a real structural resemblance of the protein binding sites. As outlined in Chapter 1, cases are known where proteins with the same fold carry out different functions and vice versa. Moreover, the binding of the same ligand is not necessarily due to a similar binding site topology, as ligands can be bound in different conformation and orientation, with proteins binding to different parts of the ligands.

Thus, it might be possible for a structure with no resemblance to the query to meet one of these criteria. Yet, since the methods presented here do measure graph similarity, encountering a high ranking graph that meets one or more of these criteria is a strong indication that the graph similarity is reflecting an actual structural similarity.

Rankings were calculated for each algorithm, including the competitor methods, based on the underlying distance/similarity measures. Retrieved items were judged relevant according the above defined rule and an averaged 11-point precision-recall plot was derived to compare the different approaches, which is shown in Fig. 5.20.

Apparently, the best results are achieved with either the SEGA approach or the

Algorithm	$P10$	R	MAP	B_{Pref}
BFPH	0.738	0.375	0.385	0.359
BFPJ	0.913	0.392	0.393	0.369
BK	0.838	0.398	0.411	0.386
CB	**0.975**	**0.526**	**0.549**	**0.519**
FFP	0.850	0.389	0.390	0.371
FPH	0.725	0.336	0.339	0.317
FPJ	0.713	0.319	0.318	0.297
GAVEO	**0.913**	**0.481**	**0.494**	**0.469**
GAVEOc	0.900	0.422	0.427	0.394
GH	0.625	0.214	0.212	0.192
RW	0.188	0.145	0.146	0.115
SEGA	**0.988**	**0.528**	**0.551**	**0.519**
SEGAHA	0.813	0.481	0.490	0.465
SP	0.125	0.158	0.160	0.131

Table 5.13: Performance of the different approaches in terms of $P10$, R, MAP and B_{Pref}.

CavBase competitor. Unfortunately, the other approaches perform less efficient on average, although the GAVEO approach shows comparably good results in the previous sections and SEGAHA is not far behind. However, all approaches, even the local fingerprints, are again able to achieve clear improvements over the baselines GH, RW and SP.

Additionally, the four evaluation criteria for retrieval systems used in the previous section ($P10$, R, MAP and B_{Pref}) were calculated, given in Table 5.13.

The performance measures largely confirm the impression given by Fig. 5.20, showing that the best performance is reached by SEGA and CB, followed by GAVEO. The higher $P10$ value of SEGA indicates that SEGA yields more consistent results on the top ranks, whereas CB has a slight advantage at higher recall levels. This is supported by the fact that the difference apparent for $P10$ and R is less pronounced for the MAP measure, which uses more information over the complete rankings. The B_{pref} measure, being more stable in case of incomplete rankings, shows no difference at all.

Apparently, the local approaches perform relatively weak. While they are still capable of outperforming the baselines SP and RW as well as the greedy approach, they fail to reach results comparable to those of SEGA or CB. Among the global approaches, GAVEO is more powerful in this scenario than GAVEOc, showing no benefit for keeping the clique solution.

Judging the results based on the above-defined rule might be incomplete or erroneous to a certain degree. As stated above, meeting these criteria is not sufficient proof for a real structural resemblance. On the other hand, if a structure has not yet been crystallized with a certain ligand, this is no indication that the ligand would be incapable of affecting the protein. In fact, the number of crystallized proteins is minimal, eclipsed by the complete set of known protein sequences. Hence, a judgment of the retrieved proteins (or items, in the context of database retrieval) might still be incomplete and most likely underestimate the percentage of items with a real similarity to the query. Therefore, the top ranks of each algorithm are also evaluated manually.

One could argue that the impression given by the precision-recall curves understate the discriminative power of the similarity measures, since the true number of relevant items is not known in advance. Assuming that unexpected similarities might be recovered, which are not known in advance, the first 100 ranks were also compared directly, adopting a simple strategy: The number of retrieved items is plotted against the number of relevant items. Thus, a perfect performance (meaning only relevant items were retrieved) will be indicated by the diagonal, whereas a suboptimal ranking will be below the diagonal.

In information retrieval, evaluation is typically restricted to the top ten or 25 ranks. However, for the above argument that interesting unexpected similarities are more likely to show up at lower ranks, the top one hundred[6] results are investigated manually, to allow for a broader view on the results. Fig. 5.21, Fig. 5.22 and Fig. 5.23 shows the results for the top 100 rankings.

The results obtained for the individual queries confirm the impression above, showing that in most cases the optimal performance is reached by CB and SEGA, while the other approaches show comparably good performance only for some of the queries.

For the first query shown in Fig. 5.21a, candidapepsin 2 from *C. albicans* (1eag), the optimal performance is achieved by SEGA, GAVEO and CB, followed by GAVEOc and SEGAHA, which show a few false positives. Unfortunately, the other approaches fail to produce similarly strong results, reaching a stagnation early on, where mostly false positives are retrieved.

The global GAVEO and GAVEOc methods are able to recover more relevant structures than the local methods, especially GAVEO, which is again superior to GAVEOc here. Among the local approaches, the best results are achieved by the fuzzy fingerprints,

[6]This value is still somewhat arbitrary, representing a compromise between feasibility and thoroughness.

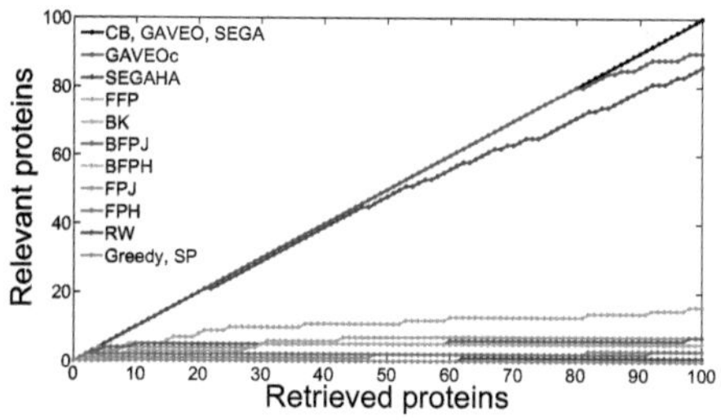

(a) 1eag

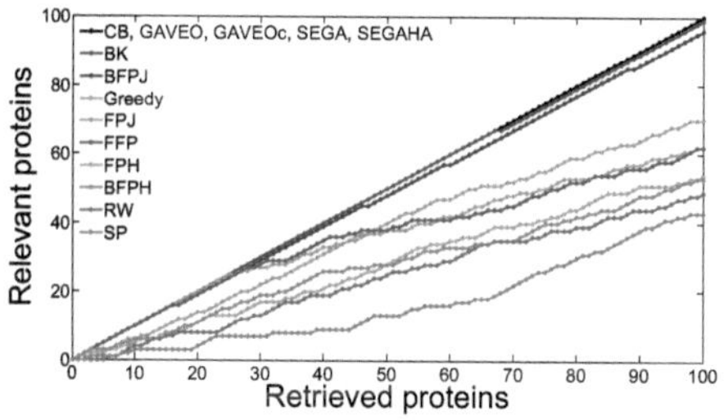

(b) 2oq5

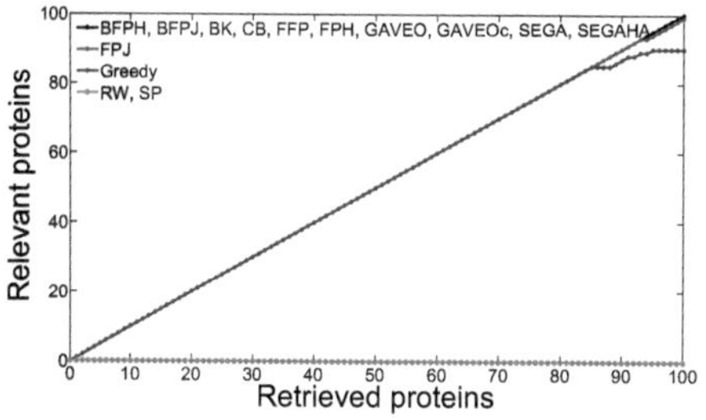

(c) 2eu2

Figure 5.21: First 100 ranks retrieved by the different similarity measures. The number of retrieved proteins is plotted against the number of relevant retrieved items.

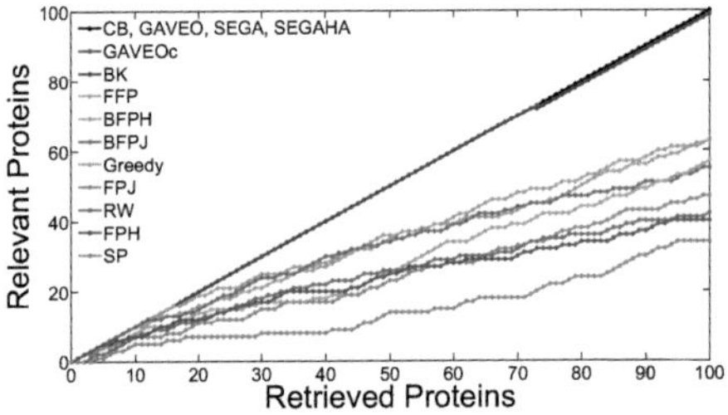

(a) 3hec

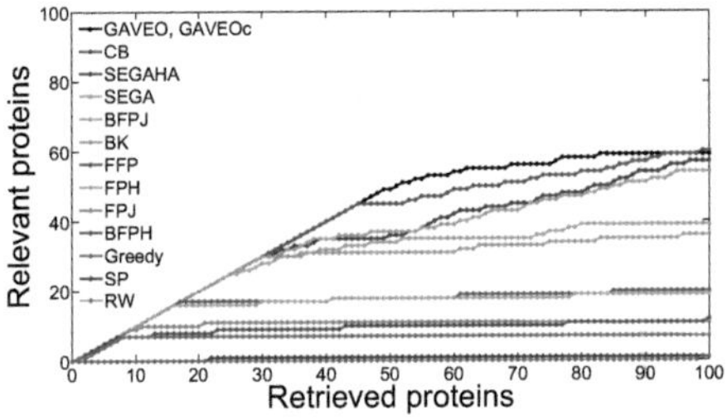

(b) 1lib

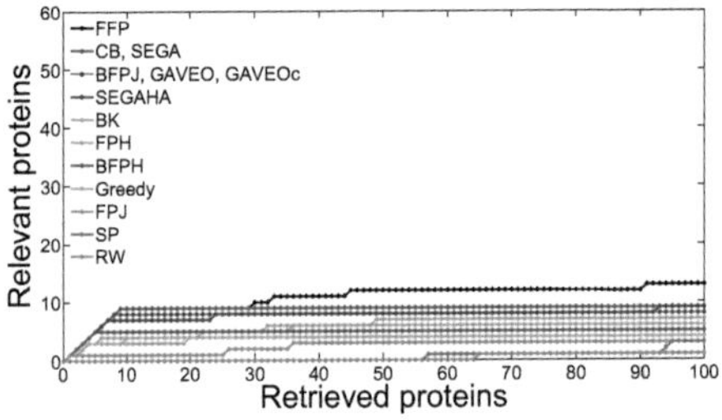

(c) 1lhu

Figure 5.22: First 100 ranks retrieved by the different similarity measures. The number of retrieved proteins is plotted against the number of relevant retrieved items.

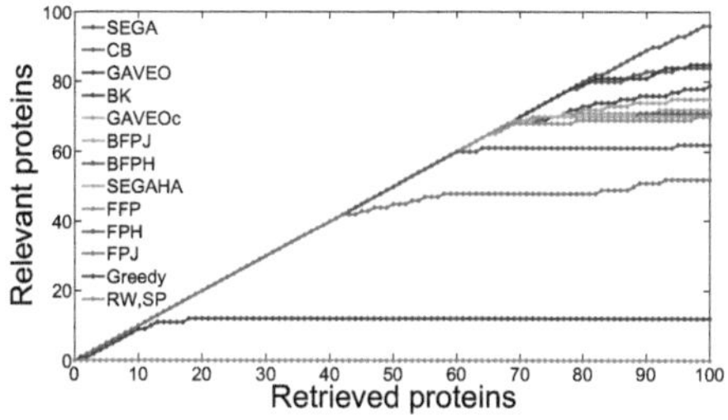

(a) 1tmn

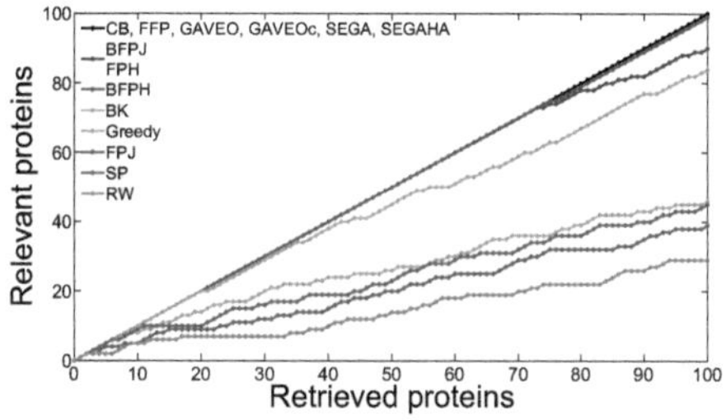

(b) 2r38

Figure 5.23: First 100 ranks retrieved by the different similarity measures. The number of retrieved proteins is plotted against the number of relevant retrieved items.

although the performance is not comparable to CB or SEGA. All of the obtained proteins that were judged relevant belong to the family of aspartic protease, which was to be expected.

Table 5.14 shows some of the obtained results for SEGA, GAVEO and CB. All three approaches retrieve similar proteins, candidapepsin being the most abundant among the top ranks. This indicates that all three are capable of retrieving relevant similarities, e.g., human β-secretase, a membrane-bound aspartic protease that cleaves the amyloid precursor protein concentrated in synapses and nerve cells in the human brain, and other aspartic proteases.

While retrieving similar proteins, the rankings produced by the CB, GAVEO and SEGA show individual differences, especially when regarding the retrieved cavities. The alignments produced by the global and semi-global approaches allow a more detailed assessment of such cases.

Rank	PDB	Function	Score	P-value
6	4er4	Endothiapepsin	30.2	n.a.
7	1e82	Endothiapepsin	30.0	n.a.
8	1e82	Endothiapepsin	29.7	n.a.
10	1e5o	Endothiapepsin	29.4	n.a.
11	1epo	Endothiapepsin	29.3	n.a.
12	1od1	Endothiapepsin	28.0	n.a.
17	1pso	Pepsin	27.2	n.a.
20	1lyb	Cathepsin D	26.5	n.a.
23	2hiz	β-Secretase	26.5	n.a.

(a) CB

Rank	PDB	Function	Score	P-value
1	1zap	Candidapepsin	2.760	$< 10^{-12}$
4	1ym4	β-Secretase	1.743	$1.3 \cdot 10^{-5}$
5	1e80	Endothiapepsin	1.688	$7.9 \cdot 10^{-5}$
7	3c9x	Aspartic protease *T.resei*	1.653	$1.9 \cdot 10^{-4}$
13	1od1	Endothiapepsin	1.627	$3.2 \cdot 10^{-4}$
35	1uh7	Rhizopepsin	1.568	$9.2 \cdot 10^{-4}$

(b) GAVEO

Rank	PDB	Function	Score	P-value
6	2er6	Endothiapepsin	8.35	$1.4 \cdot 10^{-4}$
8	5er2	Endothiapepsin	8.35	$1.4 \cdot 10^{-4}$
9	3ixj	β-Secretase	8.03	$2.1 \cdot 10^{-4}$
10	1od1	Endothiapepsin	7.83	$2.7 \cdot 10^{-4}$
19	2jjj	β-Secretase	7.24	$6.0 \cdot 10^{-4}$
20	1psn	Pepsin	7.22	$6.2 \cdot 10^{-4}$

(c) SEGA

Table 5.14: Examples of retrieved proteins for the query 1eag (secreted aspartic protease).

Fig. 5.24 shows the cavity of a human β-secretase (3ind) which was retrieved among the top 100 ranks by GAVEO but not by CB and SEGA, where it was found at ranks below 200.

It can be observed that GAVEO manages to assign the pseudocenters of catalytic residues[7] correctly to each other, as well as those of some surrounding residues. The red regions indicate similar parts of the cavities that are indeed assigned to one another. It should be noted, that SEGA manages a similar alignment, although this was rated with

[7]For each example, catalytic residues were identified according to the Catalytic Site Atlas (Porter et al., 2004).

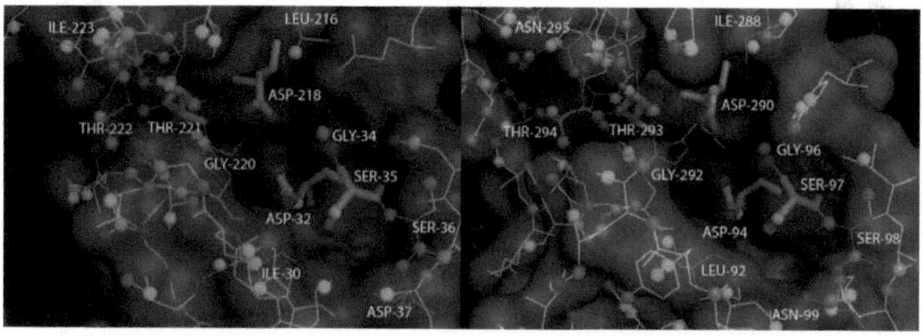

(a) Candidapepsin 2 (1eag) (b) β-Secretase (3ind)

Figure 5.24: Comparison of the main pockets of candidapepsin 2 (green) and human β-secretase 1 (cyan) as calculated by GAVEO. The red regions are assigned to each other in the corresponding graph alignment. Catalytic residues are shown in sticks representation.

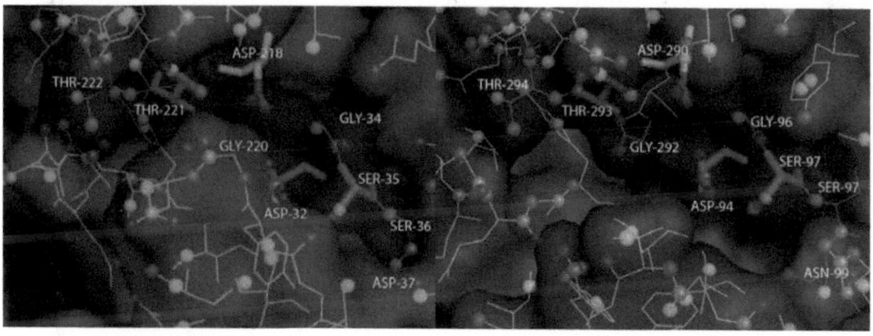

(a) Candidapepsin 2 (1eag) (b) β-Secretase (2zdz)

Figure 5.25: Comparison of the main pockets of candidapepsin 2 (green) and human β-secretase 1 (blue) as calculated by SEGA. The red regions are assigned to each other in the corresponding graph alignment. Catalytic residues are shown in sticks representation (thick lines).

a lower, yet still significant score with the SEGA distance measure.

Likewise, SEGA retrieved some cavities, that were absent in the top 100 ranks for the other approaches. Fig. 5.25 shows another cavity of human β-secretase, which was ranked well below 100 for CB and GAVEO, in the latter case even below 8000.

As can be seen, also SEGA manages to assign the pseudocenters belonging to catalytic residues successfully to one another, along with pseudocenters of some neighboring residues. The fact that GAVEO cannot retrieve this cavity at a reasonable high rank is somewhat astounding and might be due to premature termination of the optimization process. In both examples, it can be observed that the matched part is indeed structurally very similar.

For the serine protease query DESC1 (2oq5, Fig 5.21b), several approaches show a good performance, including SEGA, SEGAHA, GAVEO, GAVEOc and the baseline CB, but also the Bron-Kerbosch approach alone and one fingerprint method, BFPJ. Here, it can be observed, that the global and semi-global methods proposed in this thesis proved most successful, yielding 100 true positives on the first 100 ranks, which is only achieved by the CB baseline. All other baselines not augmented by additional surface information yield less relevant hits. Among the local methods, the BFPJ variant is most successful, though less so than the global and semi-global approaches.

Regardless of the approach, the relevant structures retrieved contained exclusively serine proteases exhibiting a trypsin-like fold (belonging to he superfamily of trypsin-like proteases according to SCOP), some of them being transmembrane serine proteases. Some examples are shown in Table 5.15. These included mainly trypsin, coagulation factor XI, hepsin, tryptase, thrombin, kallikrein and urokinase-type plasminogen activator.

CB additionally showed factor I and tryptase. The latter was also found in the GAVEO top 100 ranks. Interestingly, in case of SEGA, tryptase was only found at ranks below 100, but instead contained factor VII and X among the top ranks. Factor VII was also found among the top 100 ranks with BFPJ. Nevertheless, SEGA is able to correctly detect commonalities between tryptase and DESC 1 with a significantly high score, as can be observed in Fig. 5.26. The overlay nicely demonstrates that SEGA is able to detect a common subpocket, even so the larger part of the pockets is different.

Subtilisin was not detected at the first 100 ranks, but instead was found around rank 1,000 for the best-performing methods. Here, it should be mentioned that for CB, GAVEO, GAVEOc, SEGA and SEGAHA, the top 900 ranks almost exclusively showed serine proteases, apart from occasional false positives. With an increase in false positives, the first subtilisin-like proteases emerge.

Fig. 5.27 shows an example of a structure that was ranked high for both GAVEO and SEGA, but reached only rank 369 for CB. As can be seen, both cavities show a similar pseudocenter distribution in the pocket center, which can be correctly assigned by GAVEO. The third residue of the catalytic triad, ASP-102 was absent in case of 1lpg, hence it could not be assigned. Likewise, the second doneptor center of HIS-57 was not present in 1lpg, due to a slightly different positioning of the imidazol ring. Yet, the remaining centers of the two catalytic residues are still assigned correctly.

Rank	PDB	Function	Score	P-value
2	1qsf	Hepsin	43.9	n.a.
3	1zsl	Factor XI	43.9	n.a.
5	1o5b	u-Plasminogen activator	42.6	n.a.
6	2zdn	Trypsin	42.3	n.a.
9	2fpz	Tryptase	42.1	n.a.
15	1anw	Kallikrein	41.0	n.a.
18	1sgi	Thrombin	41.0	n.a.

(a) CB

Rank	PDB	Function	Score	P-value
2	1zrk	Factor XI	2.545	$1.2 \cdot 10^{-9}$
1	2ayw	Trypsin	2.446	$4.3 \cdot 10^{-9}$
14	2f9p	Tryptase	2.223	$9.4 \cdot 10^{-8}$
43	1o5e	Hepsin	2.151	$2.6 \cdot 10^{-7}$
44	2any	Kallikrein	1.627	$8.1 \cdot 10^{-4}$
66	1lpg	Factor X	1.568	$2.1 \cdot 10^{-3}$

(b) GAVEO

Rank	PDB	Function	Score	P-value
2	1zrk	Factor XI	24.83	$3.1 \cdot 10^{-13}$
3	1lpg	Factor X	23.79	$9.8 \cdot 10^{-13}$
5	1z8g	Hepsin	22.92	$2.6 \cdot 10^{-12}$
6	2d8w	Trypsin	22.91	$2.6 \cdot 10^{-12}$
32	2zlw	Factor VII	20.44	$4.3 \cdot 10^{-11}$
74	1o5a	u-Plasminogen activator	18.24	$5.7 \cdot 10^{-10}$

(c) SEGA

Rank	PDB	Function	Score	P-value
2	1fn8	Trypsin	851.4	$1.4 \cdot 10^{-4}$
5	1gi9	u-Plasminogen activator	827.3	$1.4 \cdot 10^{-4}$
11	2bz6	Factor VII	818.3	$2.7 \cdot 10^{-4}$

(d) BFPJ

Table 5.15: Examples of retrieved proteins for the query 1eag (secreted aspartic protease).

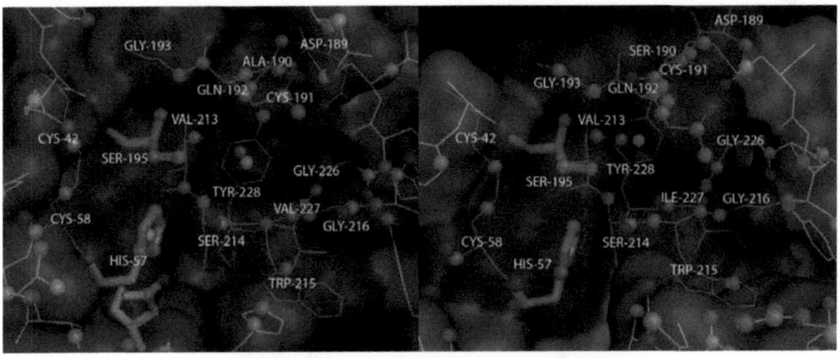

(a) DESC 1 (2oq5) (b) Tryptase (2fpz)

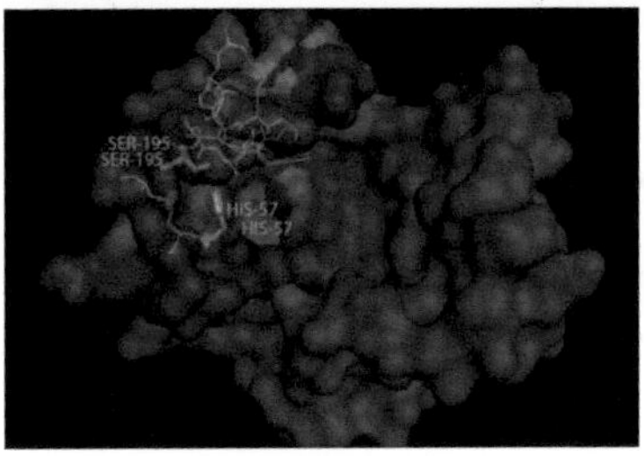

(c) Surface overlay

Figure 5.26: Comparison of the main pockets of DESC 1 (green) and human tryptase (cyan) as calculated by SEGA. The red regions are assigned to each other in the corresponding graph alignment. Catalytic residues are shown in sticks representation.

In the group of carbonic anhydrases (Fig. 5.21c), none of the approaches appear to have any difficulties, nearly all retrieving exclusively carbonic anhydrases on the top 100 ranks. The notable exception are the random walk and shortest path kernels, which fail to retrieve any meaningful result. When taking a closer look at the obtained rankings, one can find carbonic anhydrases scattered over the whole range of ranks, demonstrating that the kernel functions are not suitable as a similarity measure on protein binding sites, despite the fact that they performed well for the classification of protein structures (Borgwardt et al., 2005; Gärtner, 2003).

The greedy heuristic appears to be less powerful than the other methods, as it starts to decline earlier. This is in good agreement with the results obtained in the previous

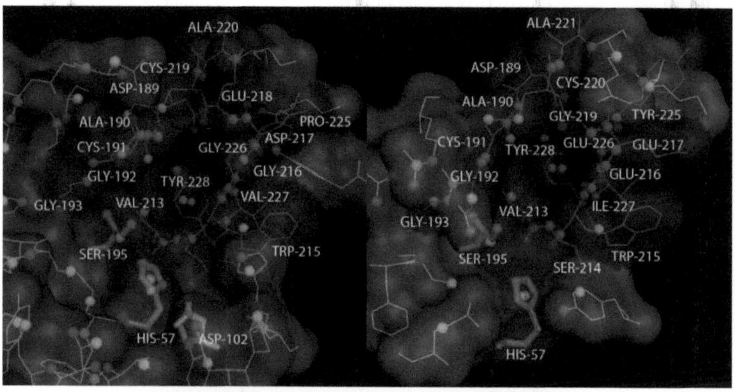

(a) DESC 1 (2oq5) (b) Factor X (1lpg)

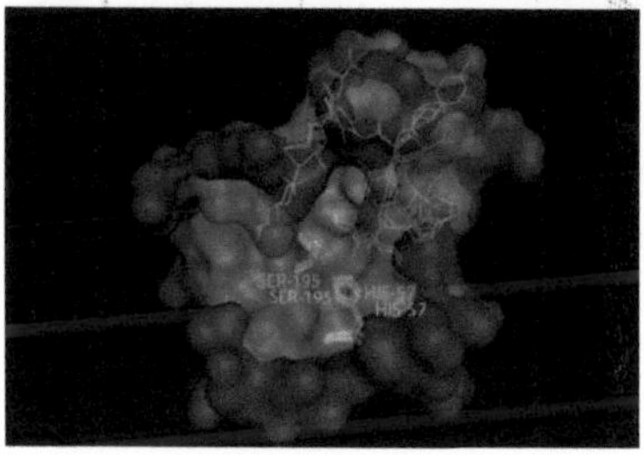

(c) Surface overlay

Figure 5.27: Comparison of the main pockets of DESC 1 (green) and human factor X (blue) as calculated by GAVEO. The red regions are assigned to each other in the corresponding graph alignment. Catalytic residues are shown in sticks representation.

sections. Apart from this, only the FPJ approach shows one false positive at rank 94. In all other cases, carbonic anhydrases (CA II) were correctly retrieved on the top ranks. In Table 5.16, the first occurrence of a protein other than carbonic anhydrase II is shown.

Generally, all of these carbonic anhydrase binding sites are globally similar in structure, an example is given in Fig. 5.28. Here, the matched region between the query and carbonic anhydrase VII is shown, as aligned by GAVEO. Not surprising, almost the complete cavity is mapped correctly to the corresponding counterpart of the query structure.

Method	Rank	PDB	Function	Score	P-value
BFPH	32	3mdz	Carbonic anhydrase VII	0.916	0.13
BFPJ	77	1jd0	Carbonic anhydrase XII	832.1	$1.7 \cdot 10^{-4}$
BK	85	1bzm	Carbonic anhydrase I	15	n.a.
CB	184	3mdz	Carbonic anhydrase VII	33.01	n.a.
FPH	104	1dmy	Carbonic anhydrase V	0.911	0.15
FPJ	101	3mdz	Carbonic anhydrase VII	4382.1	0.07
FFP	101	3mdz	Carbonic anhydrase VII	0.325	0.07
GAVEO	184	3mdz	Carbonic anhydrase VII	2.040	$1.0 \cdot 10^{-8}$
GAVEOc	150	3mdz	Carbonic anhydrase VII	2.041	$1.8 \cdot 10^{-8}$
GH	86	2ibg	Hedgehog protein	-435.6	n.a.
RW	1	1g20	Nitrogenase	$1.84 \cdot 10^{-4}$	n.a.
SEGA	150	3mdz	Carbonic anhydrase VII	24.07	$2.7 \cdot 10^{-9}$
SEGAHA	114	3mdz	Carbonic anhydrase VII	17.51	$1.0 \cdot 10^{-9}$
SP	1	3gvk	Endo-N-acetylneuraminidas	0.047	n.a.

Table 5.16: First occurrence of a protein other than carbonic anhydrase I for each approach.

For the MAP kinase query (3hec, Fig. 5.22a), several approaches achieve optimal results (CB, GAVEO, SEGA and SEGAHA), while GAVEOc and BK show only one false positive structure. Again, it can be observed that the difference between GAVEO and GAVEOc is minimal. Unfortunately, the local methods are again less successful, showing increased false positive rates, although all of them, including the kernel methods, do retrieve some relevant structures. The fuzzy fingerprints again show the best performance of the local methods.

For this query, SEGA retrieves an ephrine receptor (2qoc) with significant score. Interestingly, this structure was ranked below 1,000 by both GAVEO and CB. As can be seen, both cavities are globally dissimilar, with only a small similar subpocket, which might explain why the global methods rank this pocket rather low. The catalytic residues ASP-150 and ASP-746 are not assigned, yet, the similar subpocket around the lysine residue is mapped correctly. However, CSA indicates for this structure that the identification of the catalytic residue might be erroneous. The lysine instead is associated with ATP binding according to Uniprot (UniProt-Consortium, 2009), indicating that the matched subpocket is indeed functionally important for ATP binding.

In case of the adipocyte lipid binding protein (1lib, Fig. 5.22b), the GAVEO approach performs best, followed by CB, while SEGA and SEGAHA yield fewer relevant hits. This is in contrast to the other queries, where SEGA showed comparable or better performance

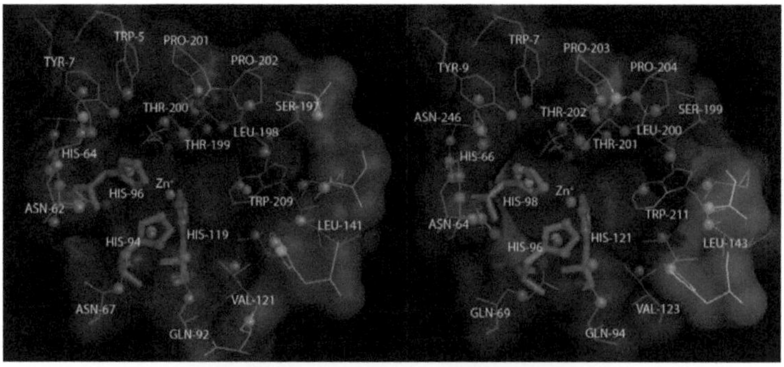

(a) Carbonic anhydrase II (2eu2) (b) Carbonic anhydrase VII (3mdz)

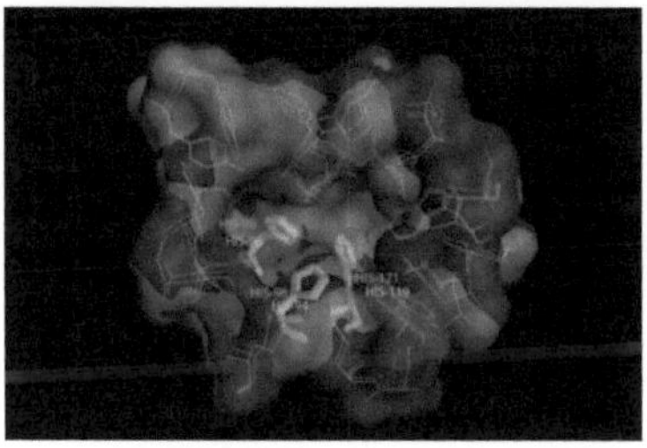

(c) Surface overlay

Figure 5.28: Comparison of the main pockets of carbonic anhydrase II (green) and carbonic anhydrase VII (yellow) as calculated by GAVEO. The red regions are assigned to each other in the corresponding graph alignment. Catalytic residues are shown in sticks representation.

than CB. In this case, the results of SEGA are even slightly worse than SEGAHA, which did not occur anywhere else.

Upon inspection of the rankings, it was noticeable, that most of the false positives corresponded to proteins associated with switchable fluorescent proteins, which also achieved significantly high scores as shown in Table 5.18. These structures where absent in other rankings, e.g., in case of the local methods or GAVEO but some of them also appeared later in the CavBase ranking.

The accumulation of these proteins in the top rank in combination with significant scores suggests that an underlying similarity exists between the query structure and the

Rank	PDB	Function	Score	P-value
2	3ctq	MAP kinase p38	71.4	n.a.
54	3cs9	Tyrosine kinase ABL1	38.7	n.a.
94	3be2	Tyrosine kinase VEGFR2	42.6	n.a.

(a) CB

Rank	PDB	Function	Score	P-value
2	3hv7	MAP kinase p38	2.690	$< 10^{-12}$
62	2ezb	Tyrosine kinase ABL1	1.820	$1.3 \cdot 10^{-5}$

(b) GAVEO

Rank	PDB	Function	Score	P-value
2	3hv6	MAP kinase p38	20.25	$1.4 \cdot 10^{-4}$
3	3mss	Tyrosine kinase ABL1	9.57	$1.4 \cdot 10^{-4}$
60	3be2	Tyrosine kinase VEGFR2	8.62	$2.1 \cdot 10^{-4}$

(c) SEGA

Table 5.17: Examples of retrieved proteins for the query 3hec (MAP kinase).

Rank	PDB	Function	Score	P-value
28	3lsa	fluorescent protein Padron0.9 (dark)	10.88	$1.2 \cdot 10^{-5}$
30	3ls3	fluorescent protein Padron0.9 (bright)	10.56	$1.7 \cdot 10^{-5}$
32	2pox	fluorescent protein Dronpa (dark)	10.27	$2.4 \cdot 10^{-5}$
33	2iov	fluorescent protein Dronpa (bright)	10.19	$2.6 \cdot 10^{-5}$
37	2gx2	fluorescent protein Dronpa	10.07	$3.0 \cdot 10^{-5}$
38	2ddc	photoswitchable fluorescent protein	10.07	$3.0 \cdot 10^{-5}$
39	2ie2	fluorescent protein Dronpa	9.93	$3.6 \cdot 10^{-5}$

Table 5.18: Examples of retrieved proteins for the query 1lib (human adipocyte lipid binding protein).

fluorescence proteins. Thus, the structures of these proteins were inspected. As it turns out, both the query structure as well as the fluorescence proteins possess a strikingly similar fold structure, consisting of a beta-sheet barrel flanked by an alpha helix. Fig. 5.30 shows the query structure compared to the first fluorescence protein (3lwa). The barrel structure is present in both cases, though with an inversed direction of the beta-strands. The cavities are obtained from the interior of the barrel. Thus, a similarity obviously does exist, which was discovered by the SEGA approach. In case of CB, similar proteins also appear, albeit on lower ranks.

The estradiol-binding site of (1lhu) has apparently only few structurally similar binding sites, indicated by the early stagnation exhibited by almost all of the approaches.

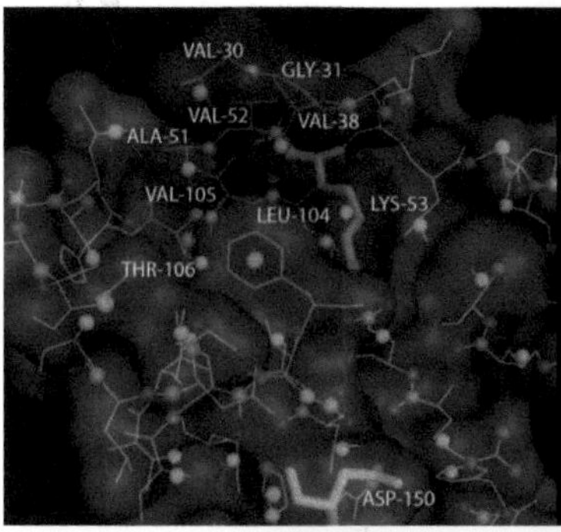

(a) MAPK 14 (3hec)

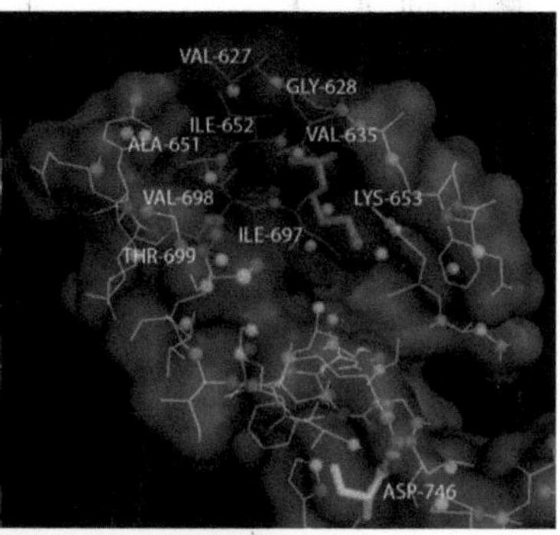

(b) EphA3 (2qoc)

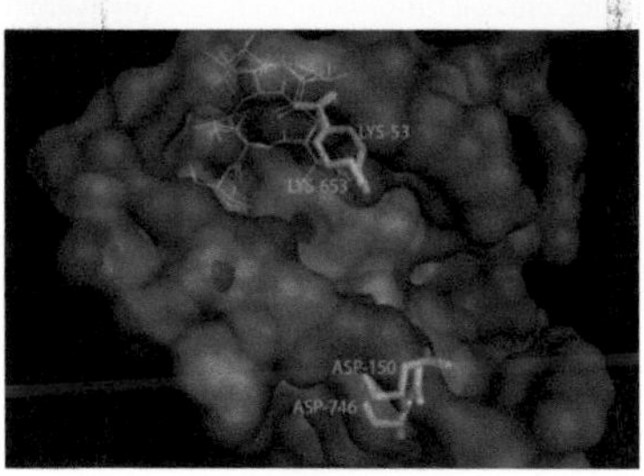

(c) Surface overlay

Figure 5.29: Comparison of the main pockets of MAPK 14 (green) and ephrine receptor EphA3 (blue) as calculated by SEGA. The red regions are assigned to each other in the corresponding graph alignment. Catalytic residues are shown in sticks representation.

BFPJ, GAVEO, GAVEOc and SEGAHA miss one or two of these structures on the top ranks (before the first false positive), though they are still retrieved among the first hundred. In total, the same nine structures where retrieved by most of the algorithms among the top ranks.

Interestingly, only the local fuzzy fingerprint approach is able to retrieve additional structures, shown in Table 5.19. These received insignificant scores from SEGA and were also ranked low by the CB. Unfortunately, despite the high rank, the scores were not significant (assuming a significance level of 0.05).

On the thermolysin query (1tmn, Fig. 5.23a), SEGA again performed best, followed

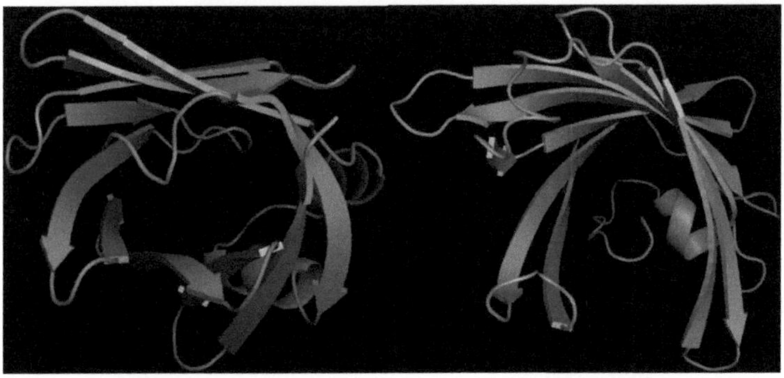

(a) adipocyte lipid binding protein(1lib) (b) Padron0.9 (3lsa)

Figure 5.30: Tertiary structures of the adipocyte lipid binding protein query (1lib) and the photoswitchable fluorescent protein Padron0.9 (3lsa).

Rank	PDB	Function	Score	P-value
30	2qxs	nuclear estradiol receptor	0.202	0.29
33	2ayr	nuclear estradiol receptor	0.201	0.30
45	2gpu	nuclear estradiol receptor	0.200	0.30
91	2p15	nuclear estradiol receptor	0.198	0.31

(a) FFP

Rank	PDB	Function	Score	P-value
12783	2gpu	nuclear estradiol receptor	5.83	n.a.
13633	2qxs	nuclear estradiol receptor	5.76	n.a.
16064	2ayr	nuclear estradiol receptor	5.54	n.a.
18330	1v0b	nuclear estradiol receptor	5.28	n.a.

(b) CB

Rank	PDB	Function	Score	P-value
14309	2qxs	nuclear estradiol receptor	4.60	0.15
24154	2ayr	nuclear estradiol receptor	3.92	0.27
19857	2P15	nuclear estradiol receptor	4.21	0.21
30462	2gpu	nuclear estradiol receptor	3.50	0.39

(c) SEGA

Table 5.19: Examples of retrieved proteins for the estradiol binding protein query (1lhu).

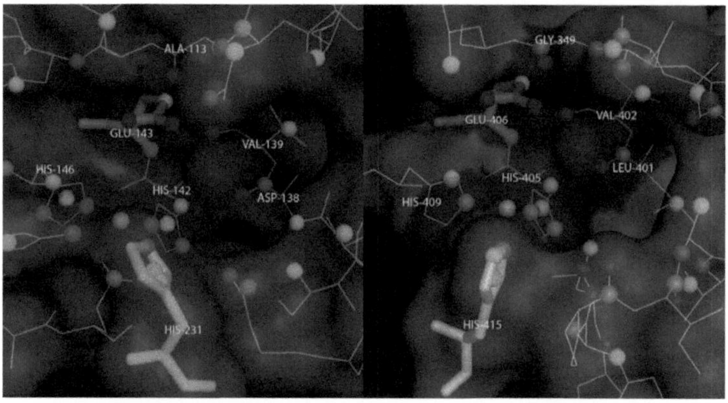

(a) Thermolysin (1tmn) (b) TACE (3g42)

Figure 5.31: Comparison of the main pockets of thermolysin (green) and TNF-alpha converting enzyme TACE (cyan) as calculated by SEGA. The red regions are assigned to each other in the corresponding graph alignment. Catalytic residues are shown in sticks representation.

by CB and GAVEO, which proves again superior to GAVEOc. The local approaches again produce a much higher false positive rate, with the fuzzy fingerprints showing the worst behavior.

Among the first one hundred ranks, SEGA retrieved several other metalloproteases besides thermolysin, some of them were also found by CavBase and GAVEO, although partly on ranks below 100. These are nevertheless related, as they can be affected by the same inhibitor. For example, elastase B from *P. aeroginosa* (pseudolysin), MMP-1 matrix metalloprotease and TACE (tumor necrosis factor-α-converting enzyme) can all be inhibited by galardin (Grobelny et al., 1992; Moss and Rasmussen, 2007), human neutral endopeptidase (neprilysin), endothelial-converting enzyme 1 and MMP-1 are both affected by thiorphan (Benchetrit et al., 1987; Roques et al., 1980; Thomas et al., 2005), both being also thermolysin inhibitors.

Fig. 5.31 shows another high ranking example obtained with SEGA. Despite the fact that the corresponding counterpart is small, SEGA manages to align this part correctly, demonstrating its capability to find also small correspondences. For comparison, this cavity was ranked below 200 by the global CB.

Rank	PDB	Function	Score	P-value
64	1z9g	thermolysin	28.1	n.a.
65	1bqb	aureolysin	26.7	n.a.
66	1r1h	neutral endopeptidase	25.9	n.a.
69	3dbk	pseudolysin	25.3	n.a.

(a) CB

Rank	PDB	Function	Score	P-value
64	1fjo	thermolysin	13.74	$9.5 \cdot 10^{-5}$
65	1ezm	pseudolysin	13.53	$1.1 \cdot 10^{-4}$
78	2qpj	neprilysin	9.52	$3.0 \cdot 10^{-4}$
84	3dwb	Endothelin-converting enzyme	9.23	$3.6 \cdot 10^{-3}$
87	3i7i	MMP-1 matrix metalloprotease	8.41	$9.4 \cdot 10^{-3}$
90	2I47	TACE	8.02	0.01

(b) SEGA

Rank	PDB	Function	Score	P-value
64	3fxs	thermolysin	1.797	$6.7 \cdot 10^{-7}$
65	3dbk	pseudolysin	1.780	$2.1 \cdot 10^{-6}$
70	3dwb	Endothelin-converting enzyme	1.585	$7.0 \cdot 10^{-4}$

(c) GAVEO

Rank	PDB	Function	Score	P-value
63	1fjw	thermolysin	0.274	0.13
64	3dbk	pseudolysin	0.262	0.15
78	2qpj	pseudolysin	0.255	0.16

(d) FFP

Table 5.20: Examples of retrieved proteins for the queries 1tmn.

Finally, on the last query (2r38, Fig. 5.23b), a structure of HIV-1 protease, the performance of SEGA and CB is rivaled by both GAVEO variants as well as SEGAHA and the fuzzy fingerprint method. All relevant retrieved structures were crystal structures of HIV-1 protease (different mutants), regardless of the method.

In total, SEGA showed the most stable performance followed closely by the CavBase approach, while GAVEO and the fuzzy fingerprints are competitive in some of the cases. Other approaches only showed a comparable high performance in the two cases where globally similar structures could be found (2eu2 and 2r38).

Regardless of the query, each of the approaches perform considerably better than the kernel methods and the greedy heuristic, which showed the worst performances in all

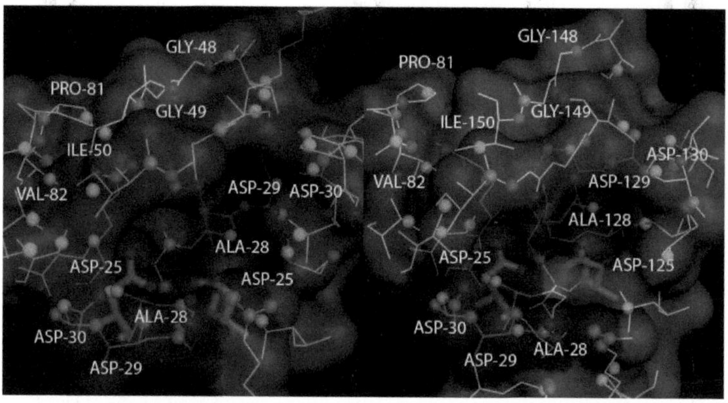

(a) HIV protease (2r38) (b) HIV protease (1nnp)

Figure 5.32: Comparison of the main pockets of two HIV protease structures by SEGA. The red and yellow regions are assigned to each other in the corresponding graph alignment. Catalytic residues are shown in sticks representation.

cases. Also noteworthy is the fact that GAVEO in all cases equals or surpasses the performance of the GAVEOc approach.

An example is again given in Fig. 5.32. In this case, SEGA matched two separate corresponding regions, demonstrating that SEGA in principle can assign multiple similar regions, even if the rest of the cavities are different.

5.7 Classification

The algorithms presented in this thesis were mainly designed to classify protein binding sites with respect to the ligand they can accommodate. Finally, the performance of the different methods for this task will be assessed. This presents yet another, more indirect way of comparing the similarity or distance measures produced by the different approaches in terms of classification accuracy, following the notion that, the more powerful a similarity measure is, the more accurate a similarity-based classifier should be when using these measures. Here, this was realized by using again a k-nearest neighbor classifier in conjunction with the different similarity scores.

k	BK	CB	GH	SA	GAVEO	GAVEO*	GAVEOc	SEGA	SEGAHA
1	76.1	81.7	76.6	89.8	89.0	78.9	87.0	91.6	89.2
3	76.3	83.1	71.8	86.8	86.5	76.6	86.5	92.4	87.9
5	77.4	83.1	72.4	86.5	86.2	78.0	86.2	91.3	85.9
7	75.7	81.1	71.8	83.7	85.4	78.6	85.6	91.6	86.8
9	76.9	79.4	76.6	82.0	85.1	76.6	85.4	92.4	84.2

(a) Global, semi-global and sequence-based approaches.

k	RW	SP	SPSA	FPJ	FPH	BFPJ	BFPH	FFP
1	59.7	60.6	62.0	87.9	80.6	89.0	87.9	83.9
3	59.7	60.6	55.2	86.5	80.3	86.2	86.8	81.7
5	59.7	63.4	55.2	84.5	77.5	84.5	85.9	79.4
7	60.8	62.5	57.2	83.6	75.8	84.2	85.4	76.6
9	60.8	63.4	57.2	80.9	73.5	83.1	84.2	75.8

(b) Local approaches.

Table 5.21: Results of k-nearest neighbor classification (percentage of correct predictions) with leave-one-out cross-validation of the original ATP/NADH dataset ($\alpha = 1$).

In addition to the algorithms employed so far, GAVEO will also be used in conjunction with the original similarity score, in order to allow for a direct comparison with the greedy heuristic. For the sake of brevity, this was excluded in the previous retrieval experiments.

5.7.1 Classification of structurally similar ATP and NADH binding sites

In Section 5.1, a first two-class dataset containing 255 proteins (15 folds) was constructed, including structures from the two most highly populated ligand groups in the CavBase: ATP and NADH. This first dataset was specifically constructed by enriching putative structurally similar binding sites within the two classes by drawing upon the spatial structure of the corresponding ligand. Hence, it is most likely to contain several largely similar binding sites.

Classification accuracy was obtained for each method from leave-one-out cross validations. As a sequence-based baseline approach, amino acid sequence alignments were calculated via the Smith-Waterman algorithm (SA). Additionally, the shortest path kernel in conjunction with sequence alignment (SPSA) is employed as a proof-of-principle. Due to its prohibitively high runtime requirement, this method is excluded from the other experiments. The obtained results are presented in Table 5.21.

When comparing the local approaches, one can see that all of the fingerprint variants are capable of considerably improving classification accuracy compared to the random walk and shortest path kernels. The most successful local method is again BFPJ, similar to the results obtained in Section 5.2.1.2. Likewise, the Jaccard measure appears to be more successful than the Hamming distance. Compared to the global approaches, one can see that all fingerprint approaches can outperform the BK baseline as well as the greedy heuristic, although only the best result, obtained by BFPJ, yields a classification accuracy that rivals the results of GAVEO and the sequence alignment baseline. The SPSA method in general appears to perform even worse than the simple SP, although for $k = 1$, the accuracy is slightly better. After all, the results do not justify the considerably larger runtime investment.

Among the global approaches, GAVEO performs slightly better than GAVEOc, indicating that a preservation of the clique solution during the optimization might not be necessary or even be detrimental, although the minor difference for $k = 1$ might simply be owed to the heuristic nature of the approaches. Both methods show results comparable to the SA baseline and outperform the greedy heuristic and the Bron-Kerbosch approach considerably. GAVEO* still achieves a higher performance than the greedy heuristic, indicating the benefit of the evolutionary optimization. Yet, the accuracy is well below the performance of GAVEO, indicating the added benefit of the modified similarity measure.

The semi-global SEGA approach can again improve upon the best local and global results as well as the sequence-based approach, while SEGAHA is comparable to those. This again indicates the benefit of drawing upon a global reference frame for resolving ambiguous matching possibilities, rather than doing so arbitrarily.

The comparably low performance of CB, as well as the inferior results of RW can partly be attributed to the fact that some comparisons could not be calculated by these approaches due to an exhaustion of runtime memory.

5.7.2 Classification of structurally diverse ATP and NADH binding sites

During the development of each algorithm, a main consideration was to account for the structural inaccuracies and variations present in structured data. Especially SEGA was developed to uncover more remote similarities than might be possible with more rigid methods. Hence, one can argue that the above used dataset is too unrealistic a test problem, since structurally similar binding sites are most likely enriched.

k	BK	CB	GH	SA	GAVEO	GAVEO*	GAVEOc	SEGA	SH
1	70.8	74.8	33.9	70.0	65.4	54.3	65.4	91.3	88.2
3	64.5	75.6	51.9	67.7	61.4	57.5	56.7	89.0	88.2
5	62.2	75.6	39.4	70.0	62.2	63.8	57.5	87.4	85.0
7	62.9	73.2	39.4	68.5	58.3	63.8	59.8	83.5	82.7
9	64.5	74.0	39.4	67.7	57.5	63.8	56.7	80.5	79.5

(a) Global, local and semi-global approaches.

k	RW	SP	FPJ	FPH	BFPJ	BFH	FFP
1	50.4	56.7	59.8	55.9	66.1	58.3	65.4
3	56.7	55.9	66.1	55.9	62.2	62.2	61.4
5	66.1	56.7	61.4	62.2	66.9	65.4	59.1
7	63.0	57.9	63.8	61.4	69.3	59.8	55.1
9	56.7	58.3	62.3	62.2	70.9	61.4	59.1

(b) Global, local and semi-global approaches.

Table 5.22: Results of k-nearest neighbor classification (percentage of correct predictions) with leave-one-out cross-validation for the one-fold ATP/NADH dataset ($\alpha = 0.1$).

For this reason, the second ATP/NADH dataset described above was compiled, which includes binding sites of proteins from different SCOP folds. More precisely, each fold was represented exactly once, as described in Section 5.1, thereby creating a rather challenging classification problem. Again, leave-one-out cross validation was performed on this two-class problem, the results are shown in Table 5.22.

As expected, all algorithms perform considerably worse on this second dataset. When comparing the local approaches, it becomes apparent that the improvement achieved by the fingerprint approaches is less obvious in this case. The best results for the local approaches are once again achieved by BFPJ. As before, the Jaccard measure proves superior to the Hamming distance. Compared to the global approaches, BFPJ performs comparable to GAVEO and GAVEOc, although the BFPJ results were achieved with a much lower runtime investment, and vastly outperforms the greedy heuristic, which performs extraordinarily weak.

Unfortunately, several competitors achieve better results than both BFPJ and GAVEO. Both the sequence-based approach as well as BK, which simply uses the size of the maximal clique as similarity measure, show a slightly better performance, and CB is even more successful. Yet, the comparably high performance of BK and CB indicates that even though proteins from different folds are used, there is still some similarity among the binding sites.

However, all of these approaches are outperformed by SEGA as well as SEGAHA, which seems in good agreement to the results obtained from the synthetic experiment, indicating that SEGA in general is more resilient towards structural differences. Again, resorting to global information seems to have a benefit, as SEGA performs better than SEGAHA.

The weak performance of the greedy heuristic is astounding at first, since BK performs much better and the Bron-Kerbosch algorithm is used to generate the seed solutions for the greedy heuristic. However, note that the BK approach uses the size of the maximal clique as a similarity measure, while the greedy heuristic uses the score of the optimization function. This supports the assumption raised in Chapter 4, that the scoring function of the greedy heuristic is not suitable as a similarity measure for protein binding sites. Still, since GAVEO* also outperforms the greedy approach, the scoring function can only partially explain the results, indicating that also the myopic nature of the heuristic plays a role.

5.7.3 Classification of the multi-class SiteEngine dataset

To test the algorithms on an independently compiled dataset, the SiteEngine dataset (cf. Section 5.1) was used again as another classification dataset. This allows the evaluation of the algorithms on a more diverse, multi-class problem setting. In total, twelve different classes are represented in the dataset. As before, leave-one-out cross validation was performed, the results are summarized in Table 5.23.

As before, the most successful local approach is BFPJ, rivaled perhaps by the fuzzy fingerprints, which seem to be more successful when confronted with more diversity in terms of different classes. Again the Jaccard coefficient generally performs better than the Hamming distance. The competitor approaches RW and SP again perform rather weak.

GAVEO and GAVEOc perform worse than BFPJ, although much better than the greedy heuristic, which again might be attributed mainly to the differing scoring functions. This is supported by the relatively strong performance of BK compared to GH. As was the case above, GAVEO* performs again better than GH, showing the benefit of the evolutionary optimization.

Apparently, both the sequence alignment and the CavBase approach achieve superior performance compared to most algorithms. Only the semi-global approaches SEGA and SEGAHA show an even better performance, arguing again in favor of a semi-global strategy. As before, SEGA improves upon the SEGAHA method.

k	BK	CB	GH	SA	GAVEO	GAVEO*	GAVEOc	SEGA	SH
1	61.7	74.8	45.4	72.2	62.1	59.1	56.0	79.8	77.0
3	54.1	69.9	33.3	68.3	58.2	55.2	53.3	76.0	72.7
5	49.2	65.0	31.7	65.9	56.0	53.6	50.0	70.5	69.4
7	44.8	61.7	30.6	65.1	54.4	54.1	50.0	68.3	69.4
9	40.4	60.1	31.1	63.4	55.5	54.6	48.4	64.5	69.4

(a) Global and semi-global approaches.

k	RW	SP	FPJ	FPH	BFPJ	BFPH	FFP
1	18.0	8.7	54.4	52.9	61.7	59.8	60.4
3	25.7	20.2	47.5	52.0	57.4	54.4	57.9
5	25.1	16.4	45.6	47.7	52.0	51.8	51.4
7	24.6	13.7	43.7	44.3	49.5	50.5	48.4
9	25.1	14.8	45.0	42.6	46.5	48.8	46.2

(b) Local approaches.

Table 5.23: Results of k-nearest neighbor classification (percentage of correct predictions) with leave-one-out cross-validation for the SiteEngine dataset ($\alpha = 1$).

5.7.4 Classification of non-native conformations on the Astex non-native dataset

Finally, to include another realistic real dataset in the analysis, the Astex non-native dataset is again utilized, this time in a classification scenario. However, unlike the previous experiments, performing an all-against-all comparison on this dataset is time-consuming and tedious, if not outright infeasible, at least without the use of high-performance computing facilities.

Therefore, instead of an all-against-all classification, the structures from the Astex diverse dataset are used as training data in order to classify the corresponding structures of the Astex non-native dataset, which serves as test dataset. The results are given in Table 5.24. Obviously, since the training set consists of only one instance per class, using a $k > 1$ would be pointless.

Again, SEGA performs best, followed closely by SEGAHA, while the GAVEO variants also achieve a high performance, outperforming all baseline approaches. Unfortunately, the fingerprint approaches prove less successful here, yielding a much lower classification accuracy. Among the local approaches, the best performance is achieved by the fuzzy fingerprints, which so far proved the most stable of the fingerprint approaches.

As can be seen, the relatively high performance of GAVEO is not at all a result of the altered scoring function, as also GAVEO* (representing GAVEO in combination with the

k	BK	CB	GH	GAVEO	GAVEO*	GAVEOc	SEGA	SEGAHA
1-NN	61.9	82.4	39.3	85.3	84.5	84.4	91.5	89.9

(a) Global and semi-global approaches.

k	BFPH	BFPJ	FFP	FPH	FPJ	RW	SP
1-NN	49.2	49.4	66.9	34.8	57.3	21.7	25.5

(b) Local approaches.

Table 5.24: Classification results on the Astex non-native dataset. Performance is measured in terms of classification accuracy.

original scoring function as similarity measure) performs equally well, while GH again performs much worse.

5.8 Virus mutants

Modeling a near-native structure solely based on sequence information is currently still an open problem, at least in the case of *ab initio* modeling, though less so for the case of homology-based methods (Kryshtafovych et al., 2009). Yet, steady improvements are being made in the field of structure prediction. As argued in Chapter 1, the advancement of such modeling techniques potentially offer applications for comparative structure and graph analysis beyond the comparison of three-dimensional protein structures and putative protein binding sites.

The purpose of the following experiment was to show, as a proof-of-principle, that this is not just a moot point but that comparative structure analysis can indeed offer real benefits for other problems as well, that are typically approached by sequence-based methods.

To this end, an HIV mutant sequence dataset compiled from the HIV Sequence Database at Los Alamos National Laboratory was used, which contained 514 mutant variants of the V3 loop of the HIV glycoprotein 120 (gp120). The dataset was used in a previous study by Sander et al. (2007) as a test dataset constituting a two-class classification problem. As mentioned in Section 5.1, each mutant strain belongs to either of the X4, R5/X4 or the R5 phenotype, indicating their capability of interacting with two different chemokine receptors. The classification problem posed in the original study was to discriminate the mutants capable of interacting with CXCR4 (X4, R5/X4) from those only capable of interacting with CCR5 (R5).

Typically, viral strains depending on CRC5 interaction to infiltrate the host cells are dominant in newly infected patients, whereas CXCR4 capable strains emerge typically in later stages of the disease, thus a co-receptor switch is assumed to be a determinant of a progression of the disease (Regoes and Bonhoeffer, 2005). Antagonists of both receptor types promise to be efficient antiviral therapeutics, though the use of CRC5 antagonists raised concerns that this might result in selective pressure towards a co-receptor switch (Westby et al., 2006). This concern along with the association of co-receptor specificity with disease progression necessitates a close monitoring of viral phenotypes.

While this can be achieved via phenotypic assays, they are relatively expensive and time-consuming. Hence, cheaper and faster methods based e.g., on sequencing techniques are desirable. A baseline for phenotype prediction based on sequence information is the 11/25 rule, which predicts a mutant to be X4-capable, if a charged amino acid can be found at position 11 or 25 of the V3 loop (De Jong et al., 1992).

This method has been improved upon by various machine learning methods (e.g., support vector machines, neural nets, etc.), among the most successful being the indicator algorithm, which uses support vector classification in combination with a linear kernel (Sing et al., 2004).

So far, the only methods also using structural descriptors have been proposed by Sander et al. (2007). $V3SD_{C\beta}$, which approximates side chains by $C\beta$ atoms and $V3SD_{SCWRL}$, which uses a more complex representation based on structure predictions using the crystal structure of 2b4c as template. Additionally, they combined structural and sequence-based features by using structural features from $V3SD_{SCWRL}$ and sequence-based features from indicator. In each case, an SVM with a Gaussian kernel was used for classification.

These methods were included here as baseline approaches, along with GH, RW and SP, which were also used in the previous experiments. To be amenable to the structure-based methods presented here, a structure model was calculated for each sequence using SCWRL with 2b4c as template structure. Sequences containing insertions or deletions were discarded. The derived structures were converted to pseudocenter representation by applying the rules of Kuhn (2005), thereby replicating the experimental setup used by Sander et al. (2007), who also employed the CavBase rules.

The algorithms used in this work so far were pitted against these baselines along with the baseline approaches used in the previous experiments, with the exception of CB and the local sequence alignment. In the former case, no stand-alone version was available that could be applied to arbitrary three-dimensional data, while the latter case was omitted, since the 11/25 rule already proved more useful. As a performance measure,

k	BFPH	BFPJ	FFP	FPH	FPJ
1	91.0	92.3	92.8	92.6	91.2
3	91.7	92.1	91.7	91.7	91.7
5	88.9	89.3	90.1	90.0	90.9
7	89.1	89.6	88.9	88.9	89.6
9	88.2	88.4	88.4	88.7	89.1

(a) local

k	GH	RW	SP
1	91.0	66.7	36.5
3	91.0	72.9	72.1
5	90.3	72.0	76.6
7	89.3	74.8	79.9
9	88.7	74.1	85.7

(b) competitor algorithms

k	GAVEO	GAVEO*	GAVEOc	SEGA	SEGAHA
1	92.4	87.5	90.1	92.6	91.9
3	91.0	88.9	90.7	90.7	90.7
5	92.1	88.9	91.2	90.3	90.0
7	90.5	87.7	89.4	90.0	90.3
9	90.7	88.0	89.6	89.1	89.8

(c) global and semi-global

Table 5.25: Results of k-nearest neighbor classification (percentage of correct predictions) with leave-one-out cross-validation for the HIV mutants dataset.

algorithm	11/25 rule	Indicator
accuracy	87.3	90.0

(a) sequence-based

algorithm	$V3SD_{C\beta}$	$V3SD_{SCWRL}$	$V3SD_{SCWRL}$ + indicator
accuracy	89.3	91.3	91.6

(b) structure-based

Table 5.26: Results of k-nearest neighbor classification (percentage of correct predictions) with leave-one-out cross-validation for the HIV mutants dataset. These are in accordance with Sander et al. (2007).

the classification accuracy was again used, derived by performing ten replicates of ten-fold stratified cross-validation.

The results obtained are shown in Table 5.25 and Table 5.26.

As becomes apparent, each of the structural approaches developed in this work is capable of improving upon the sequence-based classification of Indicator and the 11/25 rule. Interestingly, the best performance is achieved by the fingerprint approaches, notably the fuzzy fingerprint variant, though also the BFPJ and FPH variants reach high accuracies.

The semi-global SEGA and SEGAHA variants also show a high performance, with SEGAHA performing better here. Interestingly, the purely global approaches perform

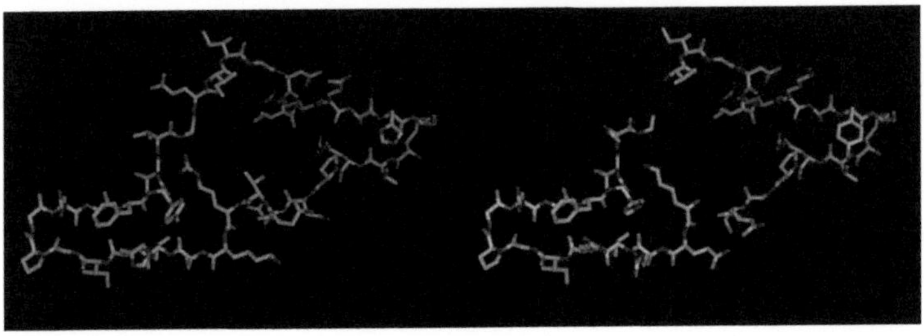

Figure 5.33: Example of 2 predicted 3D structures of the V3 loop for two different mutant strains.

slightly worse than the local and semi-global approaches.

One can argue that this is in good agreement with the fact that SEGA yields a slightly lower accuracy than SEGAHA, as it indicates that global information is less important here. Indeed, the structures obtained by the threading approach are globally rather similar, with only local differences in pseudocenter position. An example is given in Fig. 5.33. Apparently, in such cases, local methods are more useful.

Yet, each approach, including the baseline GH algorithm, yields better results than the purely sequence-based competitors, and also the more simpler $V3SD_{C\beta}$. Only the RW and SP kernels perform again much weaker, indicating their limited usefulness for such specialized problems. The fingerprint approaches (BFPJ, FPH, FFP) as well as SEGA and SEGAHA, all of them purely structural methods, moreover yield an improvement over the structural descriptors of Sander et al. (2007) and even the combination of both structural and sequence-based features as realized by the combination of $V3SD_{SCWRL}$ + Indicator, although admittedly, the difference is not large.

6

Conclusion

In this thesis, several methodically different approaches to graph comparison have been introduced for the comparative analysis of protein structure data, in particular protein binding sites. To this end, the CavBase representation was adopted. Generally speaking, the different methods can be divided into global approaches, comparing binding sites as a whole, local approaches, comparing only parts thereof, and semi-global approaches, which represent a combination of both principles.

Based on the experimental results presented in the previous chapter, one can draw some conclusions regarding the performance of the different approaches and their usefulness for the problems stated in the introduction.

6.1 Improvement of global binding site comparison

So far, the most recent global approach to protein binding site comparison was the greedy heuristic introduced by Weskamp (2007). The greedy heuristic makes use of a scoring function to evaluate the quality of a calculated graph alignment, which can intuitively be interpreted as a similarity measure on graphs and as such be used for the purpose of classifying protein binding sites and perform similarity retrieval on larger datasets given a query structure of interest.

As could be shown in the experimental chapter, this similarity measure is not well suited for the comparison of protein binding sites, since the GAVEO approach, which utilizes a modified version of this measure, produces considerably better results and supports the claim made in Chapter 4, that penalizing non-matching parts of the graph alignments will have a detrimental effect on performance.

However, as the experiments demonstrated, this increased performance is not simply a consequence of the modified similarity measure. When used in conjunction with the original scoring function, GAVEO still yielded better results than the greedy approach on average, which demonstrates the benefit of using an evolutionary strategy instead of a myopic greedy heuristic, which, in principle, allows a more thorough exploration of the search space.

Thus, one can conclude that the observed improvements can partly be attributed to a more powerful search strategy as well as to a more reasonable similarity measure. While the GAVEO approach offers the additional benefit of not being limited to graphs below a certain size, however, this comes at the price of a largely increased runtime, compared to the greedy approach.

The question whether the clique-solution should be calculated and preserved during the optimization, which was done in case of the GAVEOc variant, can generally be answered negatively, although the heuristic nature of evolutionary algorithms makes a direct comparison of the results somewhat difficult. One has to be aware of the fact that the non-deterministic approach might yield different results even for repeated runs of the same algorithm. Still, the performance difference between the variants are clear enough to be conclusive.

For the classification or retrieval experiments, the GAVEO approach showed on average a better performance than the GAVEOc variant. The main difference between the two variants is the fact that GAVEO is unrestricted in the exploration of the search space, while GAVEOc is forced to keep the clique intact. Hence, the standard GAVEO offers a bit more flexibility for the construction of alignments. Yet, GAVEOc could also yield slightly better results in some cases.

For the benchmark problems used in this work, the added flexibility might provide some advantage, as the benchmark datasets contain mostly structures with a certain degree of structural difference, in order to avoid non-trivial problems. When applied to a representative part of the complete CavBase, GAVEO also showed the most stable performance.

Yet, when retrieving structures from the complete CavBase or a sizable part thereof, the approach is confronted with a large selection of structures, increasing the chance to recover false positives. For globally more similar structures, keeping the clique solution might help to reduce the likelihood to find false positives. However, the minor differences observed in the cases, where GAVEOc yielded better results do not justify the potential further increase in runtime incurred by precalculating the clique solution. Moreover, this can only be applied if the graphs are not too large. On the other hand, starting from a good approximation of the optimal solution will in return improve the likelihood of arriving at the global optimum in fewer optimization steps.

Unfortunately, as the retrieval experiments on the high resolution set have demonstrated, the internal CavBase approach still produces better results in some scenarios, though to some extend, this might be due to a premature termination of the optimization process, which in turn might have led to unreasonable alignments. One also has to note that the CavBase approach additionally adjusts the solutions of the obtained alignments based on additional surface properties in order to remove false positives. This was not employed for the GAVEO approach, mainly with the purpose of relaxing the comparison in order to search for more remotely similar binding sites. Therefore, a rigorous filtering seemed inappropriate. As could be seen upon manually inspection, the alignments produced by GAVEO are still reasonable, mapping structurally similar regions onto each other, yet, it might be possible that using the surface information additionally might provide an edge in some cases.

Given the experimental results, which yielded a relatively high number of false positives for some queries, one can argue that including additional surface information would be beneficial for the approach, either in the form of filter approaches or directly incorporated into the optimization process. This should reduce the false positive rate for GAVEO and would also make for a more equitable comparison, still without additionally comparing surface regions. Still, the relatively high runtime demands are the most serious drawback of the GAVEO method, which precludes extensive usage on a large database.

A possible solution to this problem would be to employ faster and probably more inaccurate filtering steps in order to arrive at smaller datasets, which can then again be analyzed more thoroughly using GAVEO. Such filtering methods could in principle be realized in the form of the local fingerprint approaches introduced here. Yet, given that the local methods here are comparable in runtime requirements to SEGA, this will most likely not suffice to make GAVEO comparable to SEGA in terms of runtime.

6.2 Local methods - fast but inaccurate

Regarding the local methods, the experimental evaluation showed that the fingerprint approaches achieve a considerable improvement over the tested local kernel methods, the shortest path and the random walk kernel. These proved to be inadequate for the comparison of protein binding sites, both in terms of runtime requirements and quality of the results. As could be seen, in many cases these methods were not able to produce any meaningful result.

The astonishingly weak performance of the random walk and shortest path kernel can most likely be attributed to the principle of R-convolution kernels itself. For graphs derived from protein binding sites, the all-against-all comparison of subcomponents of structured objects appears to be unsuitable for graphs derived from protein binding pockets. Possibly, this is due to an averaging effect when comparing large graphs, or, more generally, structured objects with many components. With an increasing number of substructures, it becomes more and more likely that commonalities arise by chance.

In general, it could be observed that using the Jaccard index is in most cases superior to the Hamming distance. This is no surprise, since the Hamming distance will reward not only the presence but also the simultaneous absence of a certain pattern, which is not the case for the Jaccard measure. As argued in Chapter 4, the absence of a pattern in both structures obviously carries no information, hence the Jaccard measure would be the more appropriate measure in theory, which has proven true in the experiments.

As another observation, using a binning approach instead of ε thresholding appeared to achieve slightly better results in most cases. Especially for the classification problems, the BFPJ variant reached higher accuracies than the other variants.

Unfortunately, the fingerprint approaches could not compete with the more complex global methods, in terms of classification accuracy and retrieval performance, thus confirming the potential drawbacks of local methods raised in Section 4. By decomposing the structures into local patterns and thereby reducing the global comparison problem to a multitude of local comparisons, a loss of information is inevitably incurred. This results in a relaxation of the similarity criterion, which allows also remotely similar structures to be considered similar. On the other hand, this also increases the chance for false positives to achieve a equally high similarity as a functionally related structure.

In some cases, this inherent relaxation of the similarity criterion might even be useful. This was, for example, the case for the steroid-hormone binding globulin, for which the fuzzy fingerprints could retrieve related estradiol binding sites which other, more stringent methods could not. Yet, in most of the retrieval experiments, the number of false positives was simply too large to discern the truly related structures from similarities arising by chance, which is especially problematic for analyzing large databases.

Also the relatively weak performance on the more challenging SiteEngine and 1-fold ATP/NADH dataset can be attributed to their sensitivity to false positives, since these datasets contained much more diverse structures that the original ATP/NADH dataset. In this latter case, the fingerprint methods were much more effective.

As could be seen, this problem is even more complicated than the problem of discontinuity resulting from the harsh binning, or, respectively the ε thresholding. This is supported by the observation that the fuzzy fingerprints usually achieved the best performance in the retrieval experiments, demonstrating a real benefit of using the fuzzification to mitigate the problem of discontinuity.

However, a large number of false positives is hardly unexpected, given the local and relaxed nature of the approach. In fact, this potential drawback was already mentioned in Chapter 4. As has been stated previously, the main motivation to use local methods in the first place was to obtain a more efficient approach in terms of runtime performance. In this respect, the fingerprint approaches proved to be among the fastest methods employed in this work.

Why would this be important? As was stated in the previous section, the most severe problem of more complex methods, such as the GAVEO approach, are high runtime requirements, rendering them unsuitable for parsing a complete structural database. Instead, a fast, if maybe less accurate method can be used as a preprocessing or filtering step, limiting the number of necessary comparisons that have to be made by more time-consuming approaches. This, however, requires that the fast approach is capable of discovering real similarities among the structures to be compared.

As could be shown, this is true for the fingerprint approaches: all fingerprint variants were capable of finding related binding sites for each query structure, demonstrating that they indeed produce a measure of structural similarity. Thus, while the high false positive rates precludes a usage of the fingerprints alone for querying structural databases, they should nevertheless be useful in combination with more complex methods to speed up the retrieval process.

Moreover, the classification experiments on the modeled V3 loops of the mutant HIV-proteases showed that the fingerprint methods are more powerful on locally defined problems, where they can even yield the best performance and, more importantly, outperform the best sequence-based approaches, although this holds true for all of the structure-based methods. This experiment also indicates the potential usefulness of the fingerprint concept beyond the field of protein binding site comparison.

6.3 Combining local and global concepts

As could be seen, both global and local strategies have drawbacks as well as benefits. Thus, in order to avoid the drawbacks of both while still preserving their advantages, a semi-global methodology was suggested in this thesis, combining both concepts into a single approach. Indeed, the most versatile and efficient method appeared to be the semi-global SEGA algorithm. In most cases, SEGA indeed yielded the best results, rivaling the CavBase algorithm and sometimes outperforming it.

The only two exceptions concerned the classification of ATP and NADH binding sites on the first ATP/NADH dataset and the retrieval experiments for the adipocyte lipid binding protein. In the case of the ATP/NADH classification, the reason for the strong performance of CB can be found again in the nature of the dataset. Since both classes were constructed by deliberately enriching structures with ligands bound in similar conformation, the dataset was likely to contain structurally similar binding sites. This is obviously an advantage for the more stringent CavBase method, which, in addition to the pseudo-center representation, uses surface-based scoring schemes, in order to discard alignments where the corresponding surface patches do not show enough resemblance.

As for the adipocyte lipid binding protein, the reason for a decreased performance can be found in the structural similarity between the query and the structure of the false-positively classified fluorescent proteins. As could be shown, both structures exhibit a β-barrel, flanked by a small α helix. When limiting the comparison to the largest cavity which would correspond to the inside of the barrel, both structures are sufficiently similar to yield a statistically significant similarity score.

Compared to the other approaches, SEGA represents the best compromise between runtime efficiency and robustness. While GAVEO also proves relatively stable when confronted with noise, the runtime requirements for the evolutionary optimization are much higher, by several orders of magnitude. Yet, faster approaches are generally more intolerant towards noise and therefore more prone to return false positives, which became strikingly apparent for the fingerprint methods.

In contrast, SEGA proved much more robust towards structural variation and yet is still tolerant enough to detect similarities among related structures, even among proteins of different folds, e.g., in the case of the 1-fold ATP/NADH classification. Thus, SEGA indeed avoided the main problems of both local and global methods and instead combined their strengths.

When compared to the SEGAHA variant, representing a simpler cost-minimization approach that ignores global information, SEGA usually performs clearly better than SEGAHA. This nicely demonstrates the beneficial effect of using additional global information when constructing graph alignments instead of relying solely on local comparisons. Thus, in a sense, the SEGA approach can be considered less myopic than the purely local methods.

6.4 Global, local or semi-global?

In the context of this thesis, several conceptually and technically very different approaches to the graph alignment problem have been developed and evaluated on different problem settings. Having explored all three methodological concepts, i.e. local, global and semi-global strategies, one question remains: Which of the three principles is most suitable for the prediction of cross-reactivities and the comparison of protein binding sites?

Of course, a fundamental answer to this question that is valid across-the-board cannot be given. Instead, an answer can only be based on the premises of this work, that is, for the approaches developed here.

The experiments have shown, that each approach was based on a usable concept of similarity, since all approaches retrieved similar structures in retrieval experiments and showed classification accuracies well above random guessing, except for the R-convolution kernels. All methods were capable of identifying protein structures that can interact with or be affected by the same molecules, up to a certain degree.

The most successful method appeared to be the semi-global strategy in the form of SEGA. On the one hand, SEGA was one of the fastest and efficient approaches, on the other hand, it was also the most robust approach, being the least affected by false positive results, despite the fact that no additional surface information was used. Yet, as could be seen upon visual inspection of the cavities, the resulting alignments where nevertheless reasonable with functionally important pseudocenters mapped onto each other. Thus, in a sense, a semi-global strategy was the most appropriate in the context of this thesis. For large scale studies, the SEGA approach clearly appears to be the best choice, given its high runtime performance and the fact that it mostly showed the best results during the experiments.

Still, it would be wrong to assume that purely global or local strategies cannot be successful. In fact, the CavBase approach itself is a quite efficient global method. Despite their apparent drawbacks, the introduced global GAVEO and local fingerprint methods both have their merits and both were able to retrieve meaningful results. Still, both are also hampered by certain problems, as became apparent in the previous chapter. One possible way of alleviating these problems was already suggested above, by using the faster, but less accurate local fingerprints to perform a preselection of possibly relevant structures followed by a more thorough, albeit more time consuming comparison using a global strategy, such as GAVEO.

Such a pipeline would again, in a sense, constitute a combination of global and local methodology, thus again represent a semi-global (or semi-local) approach.

Appendix A

Data

A.1 SiteEngine dataset

Class	No.	Folds	PDB codes
Adenine-binding proteins	34	18	1a49,1a82,1ads,1atp,1ayl,1b4v,1b8a, 1bx4,1byq,1csc,1csn,1e2q,1e8x,1f9a, 1fmw,1g5t,1gn8,1hck,1hpl,1j7k,1jjv, 1kay,1kp2,1kpf,1mjh,1mmg,1nhk,1nsf, 1phk,1qmm,1yag,1zin,2src, 9ldt48
Serine proteases	24	4	1abi,1acb,1arb,1cho,1cse,1ela,1elc, 1hah,1hne,1pek,1ppf,1sbn,1sga,1sgc, 1tgs,1whs,1ysc,2alp,2lpr,3prk,3sga, 3tec,4sgb,4tgl
Fatty acid-binding	15	1	1b56,1cbs,1ftp,1hms,1ifc,1kqw,1lib, 1lid,1lie,1mdc,1opa,1opb,1pmp,2cbr, 2ifb
Estradiol-binding	11	4	1a27,1a52,1e6w,1ere,1err,1fds, 1jgl,1l2i,1lhu,1qkt,3ert
Chorismate mutases	7	1	1com,1csm,1dbf,1ecm,1fnj,1fnk,4csm2
Retinoic acid-binding	6	3	1fby,1fem,1g5y,1gx9,1tyr,2lbd
Anhydrases	6	1	1azm,1flj,1jd0,1keq,1kop,1znc
Antibiotics	6	1	1alq,1bt5,1dcs,1exm,1ghp,1rxf
HIV-1 protease	6	1	1b60,1hsg,1hsh,1hwr,1kzk,1pro
HIV-1/HIV-2	4	1	1har,1mml,1mu2,1vrt
Viral proteinase	4	1	1cqq,1lvo,1mbm,1q2w
Equilin binding proteins	3	3	1equ,1oh0,1qjg

Table A.1: SiteEngine dataset as published by (Shulman-Peleg et al., 2004)

Appendix B

Complete Results

B.1 Results from the parameter studies

		b									
		0.5	1.0	1.5	2.0	2.5	3.0	3.5	4.0	4.5	5.0
k	1	0.535	0.705	0.715	0.75	0.77	0.73	0.73	0.69	0.72	0.7
	3	0.565	0.66	0.685	0.75	0.745	0.765	0.705	0.69	0.695	0.67
	5	0.51	0.605	0.655	0.685	0.685	0.67	0.7	0.675	0.66	0.635
	7	0.52	0.565	0.615	0.625	0.68	0.65	0.71	0.645	0.625	0.64
	9	0.48	0.53	0.61	0.64	0.665	0.615	0.695	0.625	0.665	0.605

(a) Hamming distance.

		b									
		0.5	1.0	1.5	2.0	2.5	3.0	3.5	4.0	4.5	5.0
k	1	0.655	0.74	0.77	0.785	0.765	0.795	0.78	0.775	0.72	0.77
	3	0.6	0.645	0.735	0.75	0.75	0.74	0.72	0.725	0.705	0.71
	5	0.545	0.565	0.625	0.705	0.695	0.7	0.695	0.705	0.665	0.69
	7	0.5	0.575	0.635	0.68	0.665	0.695	0.69	0.675	0.675	0.645
	9	0.475	0.505	0.605	0.67	0.66	0.675	0.645	0.675	0.675	0.635

(b) Jaccard coefficient.

Table B.1: Classification rate of a k-nearest neighbor classification using the BFP approach for different bin sizes b.

		ε									
		0.1	0.2	0.3	0.4	0.5	0.6	0.7	0.8	0.9	1
k	1	0.29	0.42	0.575	0.7	0.72	0.705	0.705	0.695	0.71	0.71
	3	0.245	0.465	0.595	0.68	0.675	0.65	0.67	0.675	0.655	0.65
	5	0.26	0.415	0.57	0.595	0.585	0.59	0.59	0.565	0.595	0.55
	7	0.25	0.38	0.535	0.545	0.545	0.575	0.545	0.54	0.545	0.56
	9	0.25	0.37	0.5	0.54	0.5	0.535	0.525	0.54	0.53	0.52

		ε								
		1.1	1.2	1.3	1.4	1.5	1.6	1.7	1.8	1.9
k	1	0.71	0.725	0.735	0.745	0.735	0.705	0.685	0.685	0.72
	3	0.645	0.65	0.66	0.675	0.665	0.65	0.63	0.63	0.65
	5	0.555	0.57	0.59	0.61	0.59	0.54	0.55	0.56	0.58
	7	0.56	0.54	0.555	0.56	0.525	0.535	0.505	0.525	0.545
	9	0.515	0.5	0.51	0.515	0.495	0.515	0.52	0.52	0.515

(b) Hamming distance.

		ε									
		0.1	0.2	0.3	0.4	0.5	0.6	0.7	0.8	0.9	1
k	1	0.445	0.505	0.685	0.72	0.76	0.755	0.765	0.76	0.74	0.75
	3	0.43	0.485	0.65	0.68	0.68	0.69	0.7	0.665	0.65	0.69
	5	0.405	0.51	0.56	0.615	0.58	0.58	0.585	0.585	0.575	0.57
	7	0.41	0.52	0.53	0.535	0.545	0.55	0.57	0.56	0.565	0.56
	9	0.49	0.475	0.51	0.525	0.535	0.55	0.53	0.52	0.55	0.55

		ε								
		1.1	1.2	1.3	1.4	1.5	1.6	1.7	1.8	1.9
k	1	0.77	0.75	0.77	0.77	0.77	0.765	0.745	0.76	0.74
	3	0.695	0.675	0.68	0.68	0.65	0.69	0.665	0.67	0.69
	5	0.57	0.56	0.565	0.57	0.555	0.58	0.605	0.59	0.58
	7	0.55	0.56	0.56	0.56	0.575	0.575	0.57	0.575	0.57
	9	0.54	0.51	0.53	0.525	0.525	0.535	0.54	0.545	0.525

(d) Jaccard coefficient.

Table B.2: Classification rate of a k-nearest neighbor classification using the FP approach for different values of ε.

		η								
		0	0.5	1	1.5	2	2.5	3	3.5	4
k	1	0.25	0.685	0.76	0.81	0.81	0.81	0.79	0.825	0.805
	3	0.25	0.675	0.765	0.77	0.76	0.785	0.755	0.79	0.76
	5	0.25	0.67	0.675	0.72	0.71	0.73	0.725	0.75	0.735
	7	0.25	0.67	0.66	0.705	0.71	0.72	0.695	0.72	0.73
	9	0.25	0.63	0.645	0.69	0.675	0.7	0.675	0.715	0.67

		b			
		4.5	5	5.5	6
k	1	0.805	0.785	0.78	0.8
	3	0.765	0.765	0.775	0.77
	5	0.725	0.695	0.72	0.71
	7	0.695	0.7	0.685	0.68
	9	0.665	0.7	0.65	0.66

Table B.3: Classification rate of a k-nearest neighbor classification using the FFP approach for different η.

	$\alpha = 0$					$\alpha = 0.5$				
n_{neigh}	1	3	5	7	9	1	3	5	7	9
2	76.5	74.0	68.5	70.0	67.5	75.0	76.5	73.0	72.0	68.0
3	77.0	76.5	75.5	72.5	71.5	78.0	76.0	77.0	75.0	70.0
4	80.0	80.5	78.5	78.0	79.5	82.0	79.5	76.5	77.0	74.5
5	82.5	80.5	78.5	79.0	78.0	81.5	81.0	77.0	75.5	73.5
6	83.5	84.0	82.0	79.5	78.5	83.5	83.0	83.0	79.0	79.5
7	84.0	83.5	79.5	80.0	80.0	85.0	82.5	81.0	79.0	80.5
8	85.5	83.0	80.0	79.0	78.5	84.0	79.5	79.5	81.0	80.0
9	85.0	82.0	83.5	79.5	77.5	86.0	81.0	81.0	77.5	78.5
10	84.5	82.0	81.5	81.0	78.0	85.5	83.0	81.5	79.0	79.0
11	85.5	82.5	82.5	81.5	76.5	85.0	82.5	81.0	78.0	80.0
12	86.0	84.5	80.0	80.5	79.0	85.5	82.5	84.0	78.5	80.5
13	86.5	84.5	82.0	80.0	78.0	85.5	83.5	82.5	82.0	80.5
14	86.0	84.0	81.5	80.0	76.5	87.5	84.0	82.5	80.0	80.0
15	87.0	84.0	81.0	81.0	77.5	87.0	83.0	83.5	82.0	80.0
16	87.5	85.0	82.5	78.5	77.5	87.5	84.5	82.0	81.5	82.0

	$\alpha = 1$				
n_{neigh}	1	3	5	7	9
2	63.5	60.5	58.0	56.5	55.0
3	64.5	63.5	56.5	58.0	62.0
4	68.5	64.0	61.5	62.5	59.0
5	72.5	69.0	67.5	65.0	65.5
6	74.0	73.5	70.0	69.0	69.0
7	77.0	74.5	73.0	72.0	72.5
8	77.0	75.0	72.0	72.5	73.0
9	77.0	75.0	73.5	72.0	72.5
10	77.5	74.5	73.5	75.5	73.5
11	77.5	74.5	74.5	75.5	75.0
12	77.0	75.0	75.0	74.0	76.0
13	77.5	76.0	76.0	75.0	76.0
14	78.0	75.5	75.5	74.5	76.0
15	79.0	76.5	74.5	74.0	76.5
16	79.0	75.5	75.5	75.0	75.0

Table B.4: Classification accuracy of SEGA on the four-class dataset for different values of n_{neigh} and α.

B.2 GEDV estimates for the different approaches

	ξ	σ	μ
estimate	-0.003	0.042	0.122
confidence interval	[-0.017,0.010]	[0.041,0.043]	[0.121,0.123]

(a) BFPH

	ξ	σ	μ
estimate	0.342	780.010	973.182
confidence interval	[0.322,0.362]	[764.697,795.629]	[955.352,991.011]

(b) FPJ

	ξ	σ	μ
estimate	-0.003	0.042	0.122
confidence interval	[-0.017,0.010]	[0.041,0.043]	[0.121,0.123]

(c) FPH

	ξ	σ	μ
estimate	0.041	0.074	0.122
confidence interval	[0.026,0.057]	[0.073,0.076]	[0.121,0.124]

(d) FFP

Table B.5: GEDV parameter estimates for the score distributions of 10,000 random comparisons using the fingerprint approaches.

α	μ	σ	ξ
0.0	-0.220	0.295	0.515
0.1	-0.201	0.264	0.581
0.2	-0.182	0.233	0.647
0.3	-0.162	0.203	0.714
0.4	-0.141	0.174	0.781
0.5	-0.120	0.146	0.849
0.6	-0.098	0.120	0.916
0.7	-0.077	0.098	0.982
0.8	-0.070	0.085	1.045
0.9	-0.078	0.087	1.104
1.0	-0.091	0.099	1.157

(a) estimates

α	μ	σ	ξ
0.0	[-0.225,-0.216]	[0.291,0.300]	[0.509,0.521]
0.1	[-0.205,-0.197]	[0.260,0.268]	[0.575,0.586]
0.2	[-0.185,-0.178]	[0.230,0.237]	[0.642,0.652]
0.3	[-0.165,-0.158]	[0.200,0.206]	[0.710,0.718]
0.4	[-0.145,-0.138]	[0.171,0.176]	[0.778,0.785]
0.5	[-0.124,-0.116]	[0.144,0.148]	[0.846,0.852]
0.6	[-0.102,-0.094]	[0.118,0.121]	[0.913,0.918]
0.7	[-0.082,-0.072]	[0.096,0.099]	[0.980,0.984]
0.8	[-0.074,-0.066]	[0.084,0.086]	[1.044,1.047]
0.9	[-0.080,-0.075]	[0.086,0.088]	[1.102,1.106]
1.0	[-0.094,-0.089]	[0.098,0.100]	[1.155,1.159]

(b) confidence intervals

Table B.6: GEDV parameter estimates for different values of α using the GAVEO approach.

α	μ	σ	ξ
0.0	-0.303	0.298	0.520
0.1	-0.283	0.265	0.584
0.2	-0.261	0.234	0.650
0.3	-0.238	0.203	0.715
0.4	-0.214	0.174	0.781
0.5	-0.187	0.146	0.847
0.6	-0.158	0.120	0.913
0.7	-0.128	0.097	0.978
0.8	-0.115	0.083	1.042
0.9	-0.125	0.082	1.101
1.0	-0.146	0.093	1.156

(a) estimates

α	μ	σ	ξ
0.0	[-0.310,-0.296]	[0.293,0.302]	[0.513,0.526]
0.1	[-0.289,-0.277]	[0.261,0.269]	[0.579,0.590]
0.2	[-0.267,-0.255]	[0.231,0.237]	[0.645,0.654]
0.3	[-0.243,-0.233]	[0.201,0.206]	[0.711,0.719]
0.4	[-0.219,-0.208]	[0.171,0.177]	[0.777,0.785]
0.5	[-0.193,-0.182]	[0.144,0.148]	[0.844,0.850]
0.6	[-0.165,-0.151]	[0.118,0.121]	[0.910,0.915]
0.7	[-0.135,-0.121]	[0.096,0.099]	[0.976,0.980]
0.8	[-0.120,-0.109]	[0.082,0.084]	[1.040,1.044]
0.9	[-0.129,-0.122]	[0.081,0.083]	[1.100,1.103]
1.0	[-0.149,-0.143]	[0.092,0.094]	[1.154,1.157]

(b) confidence intervals

Table B.7: GEDV parameter estimates for different values of α using the GAVEOc approach.

α	μ	σ	ξ
0.0	-0.024	0.832	1.289
0.1	-0.001	0.811	1.468
0.2	0.016	0.815	1.639
0.3	0.026	0.836	1.801
0.4	0.034	0.867	1.955
0.5	0.041	0.905	2.103
0.6	0.048	0.947	2.244
0.7	0.055	0.994	2.379
0.8	0.063	1.043	2.510
0.9	0.070	1.095	2.637
1.0	0.079	1.149	2.761

(a) estimates

α	μ	σ	ξ
0.0	[-0.035,-0.012]	[0.819,0.845]	[1.271,1.307]
0.1	[-0.011,0.009]	[0.798,0.823]	[1.451,1.486]
0.2	[0.007, 0.024]	[0.803,0.828]	[1.621,1.656]
0.3	[0.018, 0.033]	[0.824,0.849]	[1.783,1.819]
0.4	[0.027, 0.041]	[0.854,0.880]	[1.937,1.974]
0.5	[0.034, 0.048]	[0.891,0.918]	[2.084,2.122]
0.6	[0.041, 0.055]	[0.933,0.961]	[2.224,2.264]
0.7	[0.048, 0.062]	[0.979,1.008]	[2.358,2.400]
0.8	[0.055, 0.070]	[1.028,1.059]	[2.488,2.532]
0.9	[0.063, 0.078]	[1.079,1.112]	[2.614,2.660]
1.0	[0.070, 0.087]	[1.132,1.167]	[2.736,2.785]

(b) confidence intervals

Table B.8: GEDV parameter estimates for different values of α using the SEGAHA approach.

α	μ	σ	ξ
0.0	0.008	1.206	3.282
0.1	0.005	1.117	3.136
0.2	0.004	1.032	2.987
0.3	0.003	0.952	2.834
0.4	0.002	0.877	2.674
0.5	0.003	0.811	2.508
0.6	0.004	0.754	2.332
0.7	0.006	0.711	2.142
0.8	0.010	0.690	1.935
0.9	0.015	0.706	1.708
1.0	0.013	0.774	1.464

(a) estimates

α	μ	σ	ξ
0.0	[0.007, 0.018]	[0.764, 0.785]	[1.449, 1.478]
0.1	[0.001, 0.010]	[1.102, 1.132]	[3.115, 3.157]
0.2	[0.000, 0.008]	[1.019, 1.046]	[2.968, 3.007]
0.3	[0.000, 0.006]	[0.939, 0.964]	[2.816, 2.851]
0.4	[0.000, 0.005]	[0.866, 0.889]	[2.658, 2.691]
0.5	[0.000, 0.006]	[0.800, 0.821]	[2.493, 2.523]
0.6	[0.001, 0.007]	[0.744, 0.764]	[2.318, 2.346]
0.7	[0.003, 0.009]	[0.702, 0.720]	[2.129, 2.155]
0.8	[0.006, 0.014]	[0.681, 0.699]	[1.922, 1.948]
0.9	[0.009, 0.020]	[0.696, 0.715]	[1.695, 1.721]
1.0	[0.007, 0.018]	[0.764, 0.785]	[1.449, 1.478]

(b) confidence intervals

Table B.9: GEDV parameter estimates for different values of α using the SEGA approach.

References

.C.P. Adams and V.V. Brantner. Estimating the cost of new drug development: is it really 802 million dollars? *Health Affairs*, 25(2):420–428, 2006. 7

A.V. Aho, J.E. Hopcroft, and J.D. Ullman. *The Design and Analysis of Computer Algorithms*, volume 22. Morgan Kaufmann, 1974. 29

M.A. Aizerman, E.M. Braverman, and L. Rozonoèr. Theoretical foundations of the potential function method in pattern recognition learning. *Automation and Remote Control*, 25(6):821–837, 1964. 77

T. Akutsu. A polynomial time algorithm for finding a largest common subgraph of almost trees of bounded degree. *IEICE Transactions on Fundamentals of Electronics, Communications and Computer Sciences*, 76(9):1488–1493, 1993. 30

N.N. Alexandrov and D. Fischer. Analysis of topological and nontopological structural similarities in the PDB: New examples with old structures. *Proteins*, 25(3):354–365, 1996. 17, 26

R. Allen, L. Cinque, S. Tanimoto, L. Shapiro, and D. Yasuda. A parallel algorithm for graph matching and its MasPar implementation. *IEEE Transactions on Parallel and Distributed Systems*, 8(5):490–501, 2002. ISSN 1045-9219. 34

H.A. Almohamad and S.O. Duffuaa. A linear programming approach for the weighted graph matching problem. *IEEE Transactions on Pattern Analysis and Machine Intelligence*, 15(5):522–525, 1993. ISSN 0162-8828. 33

S.F. Altschul, W. Gish, W. Miller, E.W. Myers, and D.J. Lipman. Basic local alignment search tool. *Journal of Molecular Biology*, 215(3):403–410, 1990. 14

S.F. Altschul, T.L. Madden, A.A. Schaffer, J. Zhang, Z. Zhang, W. Miller, and D.J. Lipman. Gapped BLAST and PSI-BLAST: A new generation of protein database search programs. *Nucleic Acids Research*, 25(17): 3389–3402, 1997. 2, 14

J. An, M. Totrov, and R. Abagyan. Pocketome via comprehensive identification and classification of ligand binding envelopes. *Molecular & Cellular Proteomics*, 4(6):752–761, 2005. 22

C.D. Andersson, B.Y. Chen, and A. Linusson. Mapping of ligand-binding cavities in proteins. *Proteins*, 78(6):1408–1422, 2010. 23

A. Andreeva and A.G. Murzin. Structural classification of proteins and structural genomics: new insights into protein folding and evolution. *Acta Crystallographica Section F:*

Structural Biology and Crystallization Communications, 66(10):1190–1197, 2010. ISSN 1744-3091. 2

A. Armon, D. Graur, and N. Ben-Tal. ConSurf: an algorithmic tool for the identification of functional regions in proteins by surface mapping of phylogenetic information. *Journal of Molecular Biology*, 307(1):447–463, 2001. 21

P.J. Artymiuk, A.R. Poirrette, H.M. Grindley, D.W. Rice, and P. Willett. A graph-theoretic approach to the identification of three-dimensional patterns of amino acid side-chains in protein structures. *Journal of Molecular Biology*, 243(2):327–344, 1994. 7, 19, 26, 30, 45

D. Ashlock. *Evolutionary Computation for Modeling and Optimization*. Springer, 2006. ISBN 0387221964. 63, 75

O. Bachar, D. Fischer, R. Nussinov, and H. Wolfson. A computer vision based technique for 3-D sequence-independent structural comparison of proteins. *Protein Engineering Design and Selection*, 6(3):279–287, 1993. 19

T. Bäck, D.B. Fogel, and Z. Michalewicz. *Handbook of Evolutionary Computation*. Taylor & Francis, 1997. ISBN 0750308958. 63

T. Bäck, D.B. Fogel, and Z. Michalewicz. *Evolutionary Computation 2: Advanced Algorithms and Operators*. Taylor & Francis, 2000. ISBN 0750306653. 63

S.C. Bagley and R.B. Altman. Characterizing the microenvironment surrounding protein sites. *Protein Science*, 4(4):622635, 1995. 23

A.T. Balaban, J. Brocas, and J.E. Dubois. *Chemical Applications of Graph Theory*. Academic Press, 1976. 7, 26

E. Balas and C.S. Yu. Finding a maximum clique in an arbitrary graph. *SIAM Journal on Computing*, 15:1054–1068, 1986. 30

M.T. Barakat and P.M. Dean. Molecular structure matching by simulated annealing. III. The incorporation of null correspondences into the matching problem. *Journal of Computer-Aided Molecular Design*, 5(2): 107–117, 1991. 31

J.A. Barker and J.M. Thornton. An algorithm for constraint-based structural template matching: application to 3D templates with statistical analysis. *Bioinformatics*, 19(13): 1644–1649, 2003. 19, 101

T. Bartz-Beielstein. *Experimental research in evolutionary computation: the new experimentalism*. Springer, 2006. 112

D.M. Bayada, R.W. Simpson, A.P. Johnson, and C. Laurenco. An algorithm for the multiple common subgraph problem. *Journal of Chemical Information and Computer Sciences*, 32(6):680–685, 1992. 31

M. Ben-David, O. Noivirt-Brik, A. Paz, J. Prilusky, J.L. Sussman, and Y. Levy. Assessment of CASP8 structure predictions for template free targets. *Proteins*, 77(S9):50–65, 2009. 2

T. Benchetrit, M.C. Fournié-Zaluski, and B.P. Roques. Relationship between the inhibitory potencies of thiorphan and retrothiorphan enantiomers on thermolysin and neutral endopeptidase 24.11 and their interactions with the thermolysin active site by computer modelling. *Biochemical and Biophysical Research Communications*, 147(3):1034–1040, 1987. 167

J. Berg and M. Lässig. Local graph alignment and motif search in biological networks. *PNAS*, 101(41):14689–14694, 2004. 8, 26

H.M. Berman, J. Westbrook, Z. Feng, G. Gilliland, T.N. Bhat, H. Weissig, I.N. Shindyalov, and P.E. Bourne. The protein data bank. *Nucleic Acids Research*, 28(1):235–242, 2000. 1, 2, 19, 148, 149

H.M. Berman, J.D. Westbrook, M.J. Gabanyi, W. Tao, R. Shah, A. Kouranov, T. Schwede, K. Arnold, F. Kiefer, L. Bordoli, J. Kopp, M. Podvinec, P.D. Adams, L.G. Carter, W. Minor, R. Nair, and J. La Baer. The protein structure initiative structural genomics knowledgebase. *Nucleic Acids Research*, 37 (Suppl. 1):D365–D368, 2009. ISSN 0305-1048. 2

S. Berretti, A. Del Bimbo, and E. Vicario. Efficient matching and indexing of graph models in content-based retrieval. *IEEE Transactions on Pattern Analysis and Machine Intelligence*, 23(10):1089–1105, 2002. 32

Y.E. Bessonov. On the solution of a problem on the search for the best intersection of graphs on the basis of an analysis of the projections of the subgraphs of the modular product. *Vychisl. Sistemy*, 121:3–22, 1985. 30

H.G. Beyer and H.P. Schwefel. Evolution strategies-a comprehensive introduction. *Natural Computing*, 1(1):3–52, 2002. 65, 69, 70, 71

T.A. Binkowski and A. Joachimiak. Protein functional surfaces: global shape matching and local spatial alignments of ligand binding sites. *BMC structural biology*, 8(1):45–68, 2008. 25, 59

T.A. Binkowski, L. Adamian, and J. Liang. Inferring functional relationships of proteins from local sequence and spatial surface patterns. *Journal of Molecular Biology*, 332(2): 505–526, 2003a. 22, 24, 59

T.A. Binkowski, S. Naghibzadeh, and J. Liang. CASTp: computed atlas of surface topography of proteins. *Nucleic Acids Research*, 31 (13):3352–3355, 2003b. 6, 22

C. M. Bishop. *Pattern Recognition and Machine Learning*. Springer, 2006. 7

T.L. Blundell, H. Jhoti, and C. Abell. High-throughput crystallography for lead discovery in drug design. *Nature Reviews Drug Discovery*, 1(1):45–54, 2002. ISSN 1474-1776. 1

H.J. Böckenhauer and D. Bongartz. *Algorithmic Aspects of Bioinformatics*. Springer, 2007. ISBN 3540719121. 47

H.J. Böhm. The computer program LUDI: a new method for the de novo design of enzyme inhibitors. *Journal of Computer-aided Molecular Design*, 6(1):61–78, 1992. ISSN 0920-654X. 5

C. Borgelt and M. Fiedler. Graph mining: repository vs. canonical form. In *GfKL'08: 31st Annual Conference of the German Classification Society. Proceedings*, pages 229–236, Freiburg, Germany, April 2008. 27

C. Borgelt, T. Meinl, and M. Berthold. MoSS: a program for molecular substructure mining. In *OSDM'05: First International Workshop on Open Source Data Mining: Frequent Pattern Mining Implementations. Proceedings*, pages 6–15, Chicago, USA, August 2005. 29

K.M. Borgwardt. *Graph Kernels*. PhD thesis, Ludwig-Maximilians-Universität München, 2007. 8

K.M. Borgwardt and H.P. Kriegel. Shortest-path Kernels on Graphs. In *ICDM'05: Fifth IEEE International Conference on Data Mining. Proceedings*, pages 74–81, Las Vegas, USA, July 2005. 35, 80, 107

K.M. Borgwardt, C.S. Ong, S. Schonauer, SVN Vishwanathan, A.J. Smola, and H.P. Kriegel. Protein function prediction via graph kernels. *Bioinformatics*, 21(1):i47–i56, 2005. 26, 35, 78, 82, 160

G.P. Brady and P.F.W. Stouten. Fast prediction and visualization of protein binding pockets with PASS. *Journal of Computer-Aided Molecular Design*, 14(4):383–401, 2000. 21

A. Brakoulias and R.M. Jackson. Towards a structural classification of phosphate binding sites in protein-nucleotide complexes: an automated all-against-all structural comparison using geometric matching. *Proteins*, 56(2): 250–260, 2004. 24

M. Bredel and E. Jacoby. Chemogenomics: an emerging strategy for rapid target and drug discovery. *Nature Reviews Genetics*, 5(4): 262–275, 2004. ISSN 1471-0056. 6

C. Bron and J. Kerbosch. Algorithm 457: finding all cliques of an undirected graph. *Communications of the ACM*, 16(9):575–577, 1973. 16, 19, 23, 25, 30, 55, 61, 106

R.D. Brown, G. Jones, P. Willett, and R.C. Glen. Matching two-dimensional chemical graphs using genetic algorithms. *Journal of Chemical Information and Computer Sciences*, 34 (1):63–70, 1994. 31

M. Brudno, S. Malde, A. Poliakov, C.B. Do, O. Couronne, I. Dubchak, and S. Batzoglou. Glocal alignment: finding rearrangements during alignment. *Bioinformatics*, 19(Suppl. 1):i54–i62, 2003. ISSN 1367-4803. 3

I.J. Bruno, J.C. Cole, J.P.M. Lommerse, R.S. Rowland, R. Taylor, and M.L. Verdonk. IsoStar: a library of information about nonbonded interactions. *Journal of Computer-aided Molecular Design*, 11(6):525–537, 1997. ISSN 0920-654X. 49

C. Buckley and E.M. Voorhees. Evaluating evaluation measure stability. In *SIGIR'00:*

23rd Annual International ACM SIGIR Conference on Research and Development in Information Retrieval. Proceedings., pages 33–40, Athens, Greece, July 2000. 143

C. Buckley and E.M. Voorhees. Retrieval evaluation with incomplete information. In *SIGIR'04: 27th Annual International ACM SIGIR Conference on Research and Development in Information Retrieval. Proceedings.*, pages 25–32, Sheffield, UK, July 2004. 143

H. Bunke. Error correcting graph matching: on the influence of the underlying cost function. *IEEE Transactions on Pattern Analysis and Machine Intelligence*, 21(9):917–922, 1999. 27, 32, 34

H. Bunke. Graph matching: theoretical foundations, algorithms, and applications. In *VI'00: Vision Interface Conference. Proceedings*, pages 82–88, Montréal, Canada, May 2000. 27

H. Bunke and X. Jiang. *Graph Matching and Similarity*. Kluwer Academic Publishers, 2000. 7, 27, 29

H. Bunke and BT Messmer. Recent advances in graph matching. In *Spatial Computing: Issues in Vision, Multimedia and Visualization Technologies*, pages 169–204. World Scientific, 1997. ISBN 9810229240. 31

H. Bunke and K. Shearer. A graph distance metric based on the maximal common subgraph. *Pattern Recognition Letters*, 19(3-4): 255–259, 1998. 29

H. Bunke, X. Jiang, and A. Kandel. On the minimum common supergraph of two graphs. *Computing*, 65(1):13–25, 2000. 29

B. Bustos, D. Keim, D. Saupe, T. Schreck, and D. Vranic. An experimental comparison of feature-based 3D retrieval methods. In *3DPVT'04: International Symposium on 3D Data Processing Visualization and Transmission. Proceedings.*, pages 215–222, Thessaloniki, Greece, September 2004. 142

A.A. Canutescu, A.A. Shelenkov, and R.L. Dunbrack Jr. A graph-theory algorithm for rapid protein side-chain prediction. *Protein Science*, 12(9):2001–2014, 2003. ISSN 1469-896X. 111

J.A. Capra and M. Singh. Predicting functionally important residues from sequence conservation. *Bioinformatics*, 23(15):1875–1882, 2007. ISSN 1367-4803. 22

J.A. Capra, R.A. Laskowski, J.M. Thornton, M. Singh, and T.A. Funkhouser. Predicting protein ligand binding sites by combining evolutionary sequence conservation and 3D structure. *PLoS Computational Biology*, 5(12):e1000585, 2009. 22

J.H. Chan, J.S. Hong, L.F. Kuyper, D.P. Baccanari, S.S. Joyner, R.L. Tansik, C.M. Boytos, and S.K. Rudolph. Selective inhibitors of Candida albicans dihydrofolate reductase: activity and selectivity of 5-(arylthio)-2, 4-diaminoquinazolines. *Journal of Medicinal Chemistry*, 38(18):3608–3616, 1995. 146

J.M. Chandonia and S.E. Brenner. The impact

of structural genomics: expectations and outcomes. *Science*, 311(5759):347–351, 2006. 1

B.Y. Chen and B. Honig. VASP: a volumetric analysis of surface properties yields insights into protein-ligand binding specificity. *PLoS Computational Biololgy*, 6(8): e1000881, 2010. 25

B.Y. Chen, V.Y. Fofanov, D.H. Bryant, B.D. Dodson, D.M. Kristensen, A.M. Lisewski, M. Kimmel, O. Lichtarge, and L.E. Kavraki. The MASH pipeline for protein function prediction and an algorithm for the geometric refinement of 3D motifs. *Journal of Computational Biology*, 14(6):791–816, 2007. 20

X. Chen, ZL Ji, and Y.Z. Chen. TTD: therapeutic target database. *Nucleic Acids Research*, 30(1):412–415, 2002. ISSN 0305-1048. 5

C. Chothia and A.M. Lesk. The relation between the divergence of sequence and structure in proteins. *The EMBO journal*, 5(4): 823–826, 1986. 13

W.J. Christmas, J. Kittler, and M. Petrou. Structural matching in computer vision using probabilistic relaxation. *IEEE Transactions on Pattern Analysis and Machine Intelligence*, 17(8):749–764, 1995. 33

M.L. Connolly. Solvent-accessible surfaces of proteins and nucleic acids. *Science*, 221 (4612):709–713, 1983. 23

D. Conte, P. Foggia, C. Sansone, and M. Vento. Thirty years of graph matching in pattern recognition. *International Journal of Pattern Recognition and Artificial Intelligence*, 18(3):265–298, 2004. 8, 28

S.D. Copley, W.R.P. Novak, and P.C. Babbitt. Divergence of function in the thioredoxin fold suprafamily: evidence for evolution of peroxiredoxins from a thioredoxin-like ancestor. *Biochemistry*, 43(44):13981–13995, 2004. 4, 18

P.T. Corbett, J. Leclaire, L. Vial, K.R. West, J.L. Wietor, J.K.M. Sanders, and S. Otto. Dynamic combinatorial chemistry. *Chemical Reviews*, 106(9):3652–3711, 2006. 5

L.P. Cordella, P. Foggia, C. Sansone, and M. Vento. A (sub) graph isomorphism algorithm for matching large graphs. *IEEE Transactions on Pattern Analysis and Machine Intelligence*, 26(10):1367–1372, 2004. 30

A.D.J. Cross, R.C. Wilson, and E.R. Hancock. Genetic search for structural matching. In *ECCV'96: Fourth European Conference on Computer Vision-Volume I. Proceedings*, pages 514–525, Camebridge, UK, April 1996. 33

L. David, L.R. de Beer Tjaart, T. Janet, and O. Christine. 1,000 structures and more from the MCSG. *BMC Structural Biology*, 11(2): 2, 2011. ISSN 1472-6807. 1

E.H. Davidson, J.P. Rast, P. Oliveri, A. Ransick, C. Calestani, C.H. Yuh, T. Minokawa, G. Amore, V. Hinman, C. Arenas-Mena, Otim O., Brown C.T., Livi C.B., Lee P.Y., Revilla R., Rust A.G., Pan Z., Schilstra M.J., Clarke P.J., Arnone M.I., Rowen L.,

Cameron R.A., McClay D.R., Hood L., and Bolouri H. A genomic regulatory network for development. *Science*, 295(5560):1669–1678, 2002. 26

M.O. Dayhoff, R.M. Schwartz, and B.C. Orcutt. Atlas of protein sequence and structure. *National Biomedical Research Foundation*, 5 (Suppl. 3):Dayhoff, M.O., 1978. 25

J.J. De Jong, J. Goudsmit, W. Keulen, B. Klaver, W. Krone, M. Tersmette, and A. De Ronde. Human immunodeficiency virus type 1 clones chimeric for the envelope V3 domain differ in syncytium formation and replication capacity. *Journal of Virology*, 66(2):757–765, 1992. 176

E.W. Dijkstra. A note on two problems in connexion with graphs. *Numerische Mathematik*, 1(1):269–271, 1959. 17

O. Dror, H. Benyamini, R. Nussinov, and H. Wolfson. MASS: multiple structural alignment by secondary structures. *Bioinformatics*, 19(1):i95–104, 2003. 16, 18

A.Z. Dudek, T. Arodz, and J. Galvez. Computational methods in developing quantitative structure-activity relationships (QSAR): a review. *Combinatorial Chemistry & High-Throughput Screening*, 9(3):213–228, 2006. ISSN 1386-2073. 5

A.C.M. Dumay, R.J. van der Geest, J.J. Gerbrands, E. Jansen, and J.H.C. Reiber. Consistent inexact graph matching applied to labelling coronary segments in arteriograms. In *ICPR'92: 11th IAPR International Conference on Pattern Recognition. Proceedings*, pages 439–442, The Hague, Netherlands, August 1992. 32

S.R. Eddy. Hidden Markov models. *Current Opinion in Structural Biology*, 6(3):361–365, 1996. 14

S.R. Eddy. Profile hidden Markov models. *Bioinformatics*, 14(9):755–763, 1998. 14, 24

H. Edelsbrunner, M. Facello, and J. Liang. On the definition and the construction of pockets in macromolecules. *Discrete Applied Mathematics*, 88(1-3):83–102, 1998. ISSN 0166-218X. 22

R.C. Edgar. MUSCLE: multiple sequence alignment with high accuracy and high throughput. *Nucleic Acids Research*, 32(5): 1792–1797, 2004. 2, 14, 111

A. Falicov and F.E. Cohen. A surface of minimum area metric for the structural comparison of proteins. *Journal of Molecular Biology*, 258(5):871–892, 1996. 17

M.L. Fernández and G. Valiente. A Graph distance metric combining maximum common subgraph and minimum common supergraph. *Pattern Recognition Letters*, 22(6-7): 753–758, 2001. 29

F. Ferre, G. Ausiello, A. Zanzoni, and M. Helmer-Citterich. SURFACE: a database of protein surface regions for functional annotation. *Nucleic Acids Research*, 32 (Suppl.1):240–244, 2004. 25

J.S. Fetrow and J. Skolnick. Method for prediction of protein function from sequence using the sequence-to-structure-to-function paradigm with application to glutaredoxins/thioredoxins and T~ 1 ribonucleases. *Journal of Molecular Biology*, 281(5):949–968, 1998. 20

D. Fischer, H. Wolfson, S.L. Lin, and R. Nussinov. Three-dimensional, sequence order-independent structural comparison of a serine protease against the crystallographic database reveals active site similarities: potential implications to evolution and to protein folding. *Protein Science*, 3(5):769–778, 1994. 19

R. W. Floyd. Algorithm 97: shortest path. *Communications of the ACM*, 5(6):345–345, 1962. ISSN 0001-0782. 81

T. Fober, S. Glinca, G. Klebe, and E. Hüllermeier. Superposition and alignment of labeled point clouds. *IEEE/ACM Transactions on Computational Biology and Bioinformatics*, 99:in print, 2011. ISSN 1545-5963. 48

J. Fodor and M. Roubens. *Fuzzy Preference Modelling and Multicriteria Decision Support*. Kluwer Academic Publishers, 1994. 103

H. Fröhlich, J.K. Wegner, F. Sieker, and A. Zell. Optimal assignment kernels for attributed molecular graphs. In *ICML'05: 22nd International Conference on Machine Learning. Proceedings*, volume 119, pages 225–232, Bonn, Germany, August 2005. 36

N. Funabiki and J. Kitamichi. A two-stage discrete optimization method for largest common subgraph problems. *IEICE Transactions on Information and Systems*, 82(8):1145–1153, 1999. 31

M.R. Garey and D.S. Johnson. *Computers and Intractability: A Guide to the Theory of NP-Completeness*. W.H. Freeman & Co, 1979. 29, 59

T. Gärtner. Exponential and geometric kernels for graphs. In *NIPS'02: Workshop on Unreal Data: Principles of Modeling Nonvectorial Data. Proceedings*, pages 27–31, Whistler, Canada, December 2002. 35

T. Gärtner. A survey of kernels for structured data. *SIGKKD Explorations*, 5(1):49–58, 2003. 35, 78, 106, 160

M. Gerstein and M. Levitt. Using iterative dynamic programming to obtain accurate pairwise and multiple alignments of protein structures. In *ISMB'96: 8th International Conference on Intelligent Systems for Molecular Biology. Proceedings*, volume 4, pages 59–67, La Jolla, USA, August 1996. 16, 18

M. Gerstein and M. Levitt. Comprehensive assessment of automatic structural alignment against a manual standard, the SCOP classification of proteins. *Protein Science*, 7(2): 445–456, 1998. 16, 18

J.F. Gibrat, T. Madej, and S.H. Bryant. Surprising similarities in structure comparison. *Current Opinion in Structural Biology*, 6(3): 377–385, 1996. 16, 18

F. Glaser, Y. Rosenberg, A. Kessel, T. Pupko, and N. Ben-Tal. The ConSurf-HSSP database: the mapping of evolutionary conservation among homologs onto PDB structures. *Proteins*, 58(3):610–617, 2005. 22

F. Glaser, R.J. Morris, R.J. Najmanovich, R.A. Laskowski, and J.M. Thornton. A method for localizing ligand binding pockets in protein structures. *Proteins*, 62(2):479–488, 2006. 21

N.D. Gold and R.M. Jackson. SitesBase: a database for structure-based protein-ligand binding site comparisons. *Nucleic Acids Research*, 34(Database Issue):D231–D234, 2006. ISSN 0305-1048. 24

S. Gold and A. Rangarajan. A graduated assignment algorithm for graph matching. *IEEE Transactions on Pattern Analysis and Machine Intelligence*, 18(4):377–388, 1996. 33

J. Gough and C. Chothia. SUPERFAMILY: HMMs representing all proteins of known structure. SCOP sequence searches, alignments and genome assignments. *Nucleic Acids Research*, 30(1):268–272, 2002. 15

S. Govindarajan, R. Recabarren, and R.A. Goldstein. Estimating the total number of protein folds. *Proteins*, 35(4):408–414, 1999. 2, 4, 15

A. Grant, D. Lee, and C. Orengo. Progress towards mapping the universe of protein folds. *Genome Biology*, 5(5):107–116, 2004. 2, 4, 15

L. Gregory and J. Kittler. Using graph search techniques for contextual colour retrieval. *LNCS. Structural, Syntactic, and Statistical Pattern Recognition*, 2396/2002:193–213, 2002. 32

M. Gribskov, A.D. McLachlan, and D. Eisenberg. Profile analysis: detection of distantly related proteins. *PNAS*, 84(13):4355–4358, 1987. 14

H.M. Grindley, P.J. Artymiuk, D.W. Rice, and P. Willett. Identification of tertiary structure resemblance in proteins using a maximal common subgraph isomorphism algorithm. *Journal of Molecular Biology*, 229(3): 707–721, 1993. 16, 26

D. Grobelny, L. Poncz, and R.E. Galardy. Inhibition of human skin fibroblast collagenase, thermolysin, and Pseudomonas aeruginosa elastase by peptide hydroxamic acids. *Biochemistry*, 31(31):7152–7154, 1992. 167

J.L. Gross and J. Yellen. *Graph Theory and its Applications*. CRC Press, 2006. ISBN 158488505X. 47

C. Guda, E.D. Scheeff, P.E. Bourne, and I.N. Shindyalov. A new algorithm for the alignment of multiple protein structures using Monte Carlo optimization. In *PSB'01: Pacific Symposium of Biocomputing. Proceedings*, volume 6, pages 275–286, Hawai, USA, January 2001. 16

C. Guda, S. Lu, E.D. Scheeff, P.E. Bourne, and I.N. Shindyalov. CE-MC: a multiple protein structure alignment server. *Nucleic Acids Research*, 32(Web Server Issue):W100–W103, 2004. 16, 18

A. Guerler and E.W. Knapp. Novel protein folds and their nonsequential structural analogs. *Protein Science*, 17(8):1374–1382, 2008. 16

S. Günter and H. Bunke. Self-organizing map for clustering in the graph domain. *Pattern Recognition Letters*, 23(4):405–417, 2002. 34

H.H. Guo, J. Choe, and L.A. Loeb. Protein tolerance to random amino acid change. *PNAS*, 101(25):9205, 2004. 2, 15

D. Gusfield. Efficient methods for multiple sequence alignment with guaranteed error bounds. *Bulletin of Mathematical Biology*, 55(1):141–154, 1993. ISSN 0092-8240. 47

T.R. Hagadone. Molecular substructure similarity searching: efficient retrieval in two-dimensional structure databases. *Journal of Chemical Information and Computer Sciences*, 32(5):515–521, 1992. 31

S.S. Hannenhalli and R.B. Russell. Analysis and prediction of functional sub-types from protein sequence alignments. *Journal of Molecular Biology*, 303(1):61–76, 2000. 2

H. Hark Gan, R.A. Perlow, S. Roy, J. Ko, M. Wu, J. Huang, S. Yan, A. Nicoletta, J. Vafai, D. Sun, L. Wang, J.E. Noah, S. Pasquali, and T. Schlick. Analysis of protein sequence/structure similarity relationships. *Biophysical Journal*, 83(5):2781–2791, 2002. 13

M.J. Hartshorn, M.L. Verdonk, G. Chessari, S.C. Brewerton, W.T.M. Mooij, P.N. Mortenson, and C.W. Murray. Diverse, high-quality test set for the validation of protein- ligand docking performance. *Journal of Medicinal Chemistry*, 50(4):726–741, 2007. 110

M. Hendlich, F. Rippmann, and G. Barnickel. LIGSITE: automatic and efficient detection of potential small molecule-binding sites in proteins. *Journal of Molecular Graphics and Modelling*, 15(6):359–363, 1997. 9, 22, 48

M. Hendlich, A. Bergner, J. Günther, and G. Klebe. Relibase: design and development of a database for comprehensive analysis of protein-ligand interactions. *Journal of Molecular Biology*, 326(2):607–620, 2003. ISSN 0022-2836. 108

L. Holm and J. Park. DaliLite workbench for protein structure comparison. *Bioinformatics*, 16(6):566–567, 2000. 16

L. Holm and C. Sander. Dali: a network tool for protein structure comparison. *Trends in Biochemical Sciences*, 20(11):478–480, 1995. 16

L. Holm and C. Sander. The FSSP database: fold classification based on structure-structure alignment of proteins. *Nucleic Acids Research*, 24(1):206–209, 1996. 15

J.E. Hopcroft and R.M. Karp. An $n^{5/2}$ algorithm for maximum matchings in bipartite graphs. *SIAM Journal on Computing*, 2(4):225–231, 1973. 96

J.E. Hopcroft and J.K. Wong. Linear time algorithm for isomorphism of planar graphs. In *STOC'74: Sixth Annual ACM Symposium on Theory of Computing. Proceedings*, pages 172–184, Seattle, Washington, July 1974. 29

J. Huan, W. Wang, and J. Prins. Efficient mining of frequent subgraphs in the presence of isomorphism. In *ICDM'03: Third IEEE International Conference on Data Mining. Proceedings*, pages 549–561, Melbourne, USA, November 2003. 29

C. Huang, M. Tang, M.Y. Zhang, S. Majeed, E. Montabana, R.L. Stanfield, D.S. Dimitrov, B. Korber, J. Sodroski, I.A. Wilson, R. Wyatt, and P.D. Kwong. Structure of a v3-containing hiv-1 gp120 core. *Science*, 310 (5750):1025–1028, 2005. 111

B. Huet and E.R. Hancock. Shape recognition from large image libraries by inexact graph matching. *Pattern Recognition Letters*, 20 (11-13):1259–1269, 1999. ISSN 0167-8655. 33

N. Hulo, A. Bairoch, V. Bulliard, L. Cerutti, E. De Castro, P.S. Langendijk-Genevaux, M. Pagni, and C.J.A. Sigrist. The PROSITE database. *Nucleic Acids Research*, 34(Suppl. 1):D227–D230, 2006. ISSN 0305-1048. 25

D.H. Huson and D. Bryant. Application of phylogenetic networks in evolutionary studies. *Molecular Biology and Evolution*, 23(2): 254–267, 2006. 26

C. Irniger and H. Bunke. Graph matching: filtering large databases of graphs using decision trees. In *GbR'01: 3rd IAPR-TC15 Workshop on Graph-based Representation in Pattern Recognition. Proceedings*, pages 239–249, Ischia, Italy, May 2001. 31

V.A. Ivanisenko, S.S. Pintus, D.A. Grigorovich, and N.A. Kolchanov. PDBSiteScan: a program for searching for active, binding and posttranslational modification sites in the 3D structures of proteins. *Nucleic Acids Research*, 32(Web Server Issue):W549–W554, 2004. 20

A. Jagota, M. Pelillo, and A. Rangarajan. A new deterministic annealing algorithm for maximum clique. In *IJCNN'00: International Joint Conference on Neural Networks. Proceedings.*, pages 505–508, Como, Italy, July 2000. 33

B.J. Jain, P. Geibel, and F. Wysotzki. SVM learning with the Schur-Hadamard inner product for graphs. *Neurocomputing*, 64:93–105, 2005. 36

M. Jambon, A. Imberty, G. Deleage, and C. Geourjon. A new bioinformatic approach to detect common 3D sites in protein structures. *Proteins*, 52(2):137–145, 2003. 19, 26

M. Jambon, O. Andrieu, C. Combet, G. Deleage, F. Delfaud, and C. Geourjon. The SuMo server: 3D search for protein functional sites. *Bioinformatics*, 21(20):3929–3930, 2005. 19

L. Jaroszewski, A. Godzik, and L. Rychlewski. Improving the quality of twilight-zone alignments. *Protein Science*, 9(8):1487–1496, 2000. ISSN 1469-896X. 18

L.J. Jensen, R. Gupta, H.H. Staerfeldt, and S. Brunak. Prediction of human protein function according to Gene Ontology categories. *Bioinformatics*, 19(5):635–642, 2003. 2

M. Jost, G. Zocher, S. Tarcz, M. Matuschek, X. Xie, S.M. Li, and T. Stehle. Structure-function analysis of an enzymatic prenyl transfer reaction identifies a reaction chamber with modifiable specificity. *Journal of the American Chemical Society*, 132(50):1788–1798, 2010. ISSN 0002-7863. 3

D. Justice and A. Hero. A binary linear programming formulation of the graph edit distance. *IEEE Transactions on Pattern Analysis and Machine Intelligence*, 28(8):1200–1214, 2006. 27, 33

W. Kabsch. A solution of the best rotation to relate two sets of vectors. *Acta Crystallographica*, 32(8):922–923, 1976. 109

M. Kanehisa, S. Goto, S. Kawashima, Y. Okuno, and M. Hattori. The KEGG resource for deciphering the genome. *Nucleic Acids Research*, 32(90001):277–280, 2004. 8, 26, 27

R.M. Karp. Reducibility among combinatorial problems. In M. Jünger, T.M. Liebling, G.L. Naddef, D. Nemhauser, W.R. Pulleyblank, G. Reinelt, G. Rinaldi, and L.A. Wolsey, editors, *50 Years of Integer Programming 1958-2008*, chapter 8, pages 219–241. Springer, 2010. 59

M. Karplus and J.A. McCammon. Molecular dynamics simulations of biomolecules. *Nature Structural & Molecular Biology*, 9(9): 646–652, 2002. 5

H. Kashima, K. Tsuda, and A. Inokuchi. Marginalized kernels between labeled graphs. In *ICML'03: 20th International Conference on Machine Learning. Proceedings*, pages 321–328, Washington, DC, USA, August 2003. 35

H. Kashima, K. Tsuda, and A. Inokuchi. Kernels for graphs. In *Kernel Methods in Computational Biology*, chapter 7, pages 155–170. MIT Press, 2004. 35

K. Katoh, K. Misawa, K. Kuma, and T. Miyata. MAFFT: a novel method for rapid multiple sequence alignment based on fast Fourier transform. *Nucleic Acids Research*, 30(14): 3059–3066, 2002. 14

T. Kawabata and K. Nishikawa. Protein structure comparison using the Markov transition model of evolution. *Proteins*, 41(1):108–122, 2000. 17, 18

K. Kinoshita and H. Nakamura. Identification of protein biochemical functions by similarity search using the molecular surface database eF-site. *Protein Science*, 12(8): 1589–1595, 2003. 6, 7, 23, 26

K. Kinoshita and H. Nakamura. Identification of the ligand binding sites on the molecular surface of proteins. *Protein Science*, 14(3): 711–718, 2005. 23, 26, 30, 59, 106

J. Kittler and E.R. Hancock. Combining evidence in probabilistic relaxation. *International Journal of Pattern Recognition and Artificial Intelligence*, 3(1):29–51, 1989. 33

G. Klebe. *Wirkstoffdesign: Entwurf und Wirkung von Arzneistoffen*. Springer, 2009. ISBN 3827420466. 1, 6, 49, 148

G.J. Kleywegt. Recognition of spatial motifs in protein structures. *Journal of Molecular Biology*, 285(4):1887–1897, 1999. 20

J. Köbler, U. Schöning, and J. Torán. *The Graph Isomorphism Problem: Its Structural Complexity*. Birkhauser Verlag, 1994. 29

I. Koch. Enumerating all connected maximal common subgraphs in two graphs. *Theoretical Computer Science*, 250(1-2):1–30, 2001. 30

B. Kolbeck, P. May, T. Schmidt-Goenner, T. Steinke, and E.W. Knapp. Connectivity independent protein-structure alignment: a hierarchical approach. *BMC Bioinformatics*, 7 (1):510–530, 2006. 16, 18

R. Kondor and K.M. Borgwardt. The skew spectrum of graphs. In *ICML'08: 25th International Conference on Machine Learning. Proceedings*, pages 496–503, Helsinki, Finnland, July 2008. 36

R.I. Kondor and J. Lafferty. Diffusion kernels on graphs and other discrete structures. In *ICML'02: 19th International Conference on Machine Learning. Proceedings*, pages 315–322, Sydney, Australia, July 2002. 35

S. Kosinov and T. Caelli. Inexact multisubgraph matching using graph eigenspace and clustering models. *LNCS. Structural, Syntactic, and Statistical Pattern Recognition*, 2396/2002: 455–473, 2009. 36

E. Krissinel and K. Henrick. Secondary-structure matching (SSM), a new tool for fast protein structure alignment in three dimensions. *Acta Crystallographica Section D: Biological Crystallography*, 60(1:12):2256–2268, 2004a. 16, 18

E.B. Krissinel and K. Henrick. Common subgraph isomorphism detection by backtracking search. *Software Practice and Experience*, 34(6):591–607, 2004b. 16

A. Krogh and M. Brown. Hidden Markov models in computational biology. *Journal of Molecular Biology*, 235(2):1501–1531, 1994. 14

A. Kryshtafovych, K. Fidelis, and J. Moult. CASP8 results in context of previous experiments. *Proteins: Structure, Function, and Bioinformatics*, 77(S9):217–228, 2009. ISSN 1097-0134. 2, 175

D. Kuhn, N. Weskamp, S. Schmitt, E. Hüllermeier, and G. Klebe. From the similarity analysis of protein cavities to the functional classification of protein families using Cavbase. *Journal of Molecular Biology*, 359 (4):1023–1044, 2006. 9, 50, 51, 111

H.W. Kuhn. The hungarian method for the assignment problem. *Naval Research Logistics*, 52(1):7–21, 2005. 92, 95, 101, 106, 176

M. Kuramochi and G. Karypis. Discovering frequent geometric subgraphs. *Information Systems*, 32(8):1101–1120, 2007. 27, 29

J. Lafferty and G. Lebanon. Information diffusion kernels. *Advances in Neural Information Processing Systems*, 500(2):391–398, 2003. 35

Y. Lamdan and H.J. Wolfson. Geometric hashing: a general and efficient model-based recognition scheme. In *ICCV'88: 2nd International Conference on Computer Vision. Proceedings*, pages 238–249, Tampa, USA, December 1988. 19, 32

Y. Lamdan, J.T. Schwartz, and H.J. Wolfson. On recognition of 3D objects from 2D images. In *ICRA'88: 1988 IEEE International Conference on Robotics and Automation. Proceedings*, pages 1407–1413, Philadelphia, USA, April 1988. 19

T. Langer and R.D. Hoffmann. *Pharmacophores And Pharmacophore Searches*. Vch Verlagsgesellschaft Mbh, 2006. ISBN 3527312501. 5, 7

M.A. Larkin, G. Blackshields, N.P. Brown, R. Chenna, P.A. McGettigan, H. McWilliam, F. Valentin, I.M. Wallace, A. Wilm, R. Lopez, J.D. Thompson, T.J. Gibson, and D.G. Higgins. Clustal W and Clustal X version 2.0. *Bioinformatics*, 23(21): 2947–2948, 2007. 2, 14, 47

J. Larrosa and G. Valiente. Constraint satisfaction algorithms for graph pattern matching. *Mathematical Structures in Computer Science*, 12(4):403–422, 2002. ISSN 0960-1295. 30

R.A. Laskowski. SURFNET: a program for visualizing molecular surfaces, cavities, and intermolecular interactions. *Journal of Molecular Graphics*, 13(5):323–330, 1995. 21

R.A. Laskowski, N.M. Luscombe, M.B. Swindells, and J.M. Thornton. Protein clefts in molecular recognition and function. *Protein Science*, 5(12):2438–2452, 1996. 6, 21, 61

R.H. Lathrop. The protein threading problem with sequence amino acid interaction preferences is NP-complete. *Protein Engineering Design and Selection*, 7(9):1059–1068, 1994. 16

A.T.R. Laurie and R.M. Jackson. Q-SiteFinder: an energy-based method for the prediction of protein-ligand binding sites. *Bioinformatics*, 21(9):1908–1916, 2005. 22

M. Lazarescu, H. Bunke, and S. Venkatesh. Graph matching: fast candidate elimination using machine learning techniques. *LNCS. Advances in Pattern Recognition*, 1876/2000: 236–245, 2000. 31

D. Lee, O. Redfern, and C. Orengo. Predicting protein function from sequence and structure. *Nature Reviews Molecular Cell Biology*, 8(12):995–1005, 2007. 2, 14

N. Leibowitz, R. Nussinov, and H.J. Wolfson. MUSTA - a general, efficient, automated method for multiple structure alignment and detection of common motifs: application to proteins. *Journal of Computational Biology*, 8(2):93–121, 2001. 17, 18

H. Leonov, J.S.B. Mitchell, and I.T. Arkin. Monte Carlo estimation of the number of possible protein folds: effects of sampling bias and folds distributions. *Proteins*, 51(3): 352–359, 2003. 2, 4, 15

V.I. Levenshtein. Binary codes capable of

correcting deletions, insertions, and reversals. *Soviet Physics Doklady*, 10(8):707–710, 1966. 82

G. Levi. A note on the derivation of maximal common subgraphs of two directed or undirected graphs. *Calcolo*, 9(1972):341–352, 1973. 44

D.G. Levitt and L.J. Banaszak. POCKET: A computer graphies method for identifying and displaying protein cavities and their surrounding amino acids. *Journal of Molecular Graphics*, 10(4):229–234, 1992. 21

J. Liang, H. Edelsbrunner, P. Fu, P.V. Sudhakar, and S. Subramaniam. Analytical shape computing of macromolecules II: identification and computation of inaccessible cavities inside proteins. *Proteins*, 33(1):18–29, 1998a. 22

J. Liang, H. Edelsbrunner, and C. Woodward. Anatomy of protein pockets and cavities: measurement of binding site geometry and implications for ligand design. *Protein Science*, 7(9):1884–1897, 1998b. ISSN 0961-8368. 22

C.A. Lipinski, F. Lombardo, B.W. Dominy, and P.J. Feeney. Experimental and computational approaches to estimate solubility and permeability in drug discovery and development settings. *Advanced Drug Delivery Reviews*, 23(1-3):3–25, 1997. ISSN 0169-409X. 6

X. Liu, K. Fan, and W. Wang. The number of protein folds and their distribution over families in nature. *Proteins*, 54(3):491–499, 2004. 2, 4, 15

T. Madej, J.F. Gibrat, and S.H. Bryant. Threading a database of protein cores. *Proteins*, 23 (3):356–369, 1995. 16, 18, 26

P. Mahé, N. Ueda, T. Akutsu, J.L. Perret, and J.P. Vert. Extensions of marginalized graph kernels. In *ICML'04: 21st International Conference on Machine Learning. Proceedings*, pages 70–78, Banff, Canada, July 2004. 35

C.D. Manning, P. Raghavan, and H. Sch "utze. *Introduction to Information Retrieval*, volume 1. Cambridge University Press Cambridge, 2008. 138, 142, 143

M.A. Marti-Renom, E. Capriotti, I.N. Shindyalov, and P.E. Bourne. Structure comparison and alignment. In *Structural Bioinformatics*, chapter 16, pages 397–418. John Whiley & Sons, 2009. 7

K. Mason, N.M. Patel, A. Ledel, C.C. Moallemi, and E.A. Wintner. Mapping protein pockets through their potential small-molecule binding volumes: QSCD applied to biological protein structures. *Journal of Computer-aided Molecular Design*, 18(1): 55–70, 2004. 23

I.K. McDonald and J.M. Thornton. Satisfying hydrogen bonding potential in proteins. *Journal of Molecular Biology*, 238(5):777–793, 1994. ISSN 0022-2836. 49

J.J. McGregor. Backtrack search algorithms and the maximal common subgraph problem. *Software - Practice and Experience*, 12(1): 23–34, 1982. 30

B.D. McKay. *Practical Graph Isomorphism.* Citeseer, 1981. 31

S. Medasani, R. Krishnapuram, and Y.S. Choi. Graph matching by relaxation of fuzzy assignments. *IEEE Transactions on Fuzzy Systems*, 9(1):173–182, 2002. ISSN 1063-6706. 33

B.T. Messmer and H. Bunke. Error-correcting graph isomorphism using decision trees. *International Journal of Pattern Recognition and Artificial Intelligence*, 12(6):721–742, 1998. 33

B.T. Messmer and H. Bunke. A decision tree approach to graph and subgraph isomorphism detection. *Pattern Recognition*, 32 (12):1979–1998, 1999. 31

B.T. Messmer and H. Bunke. Efficient subgraph isomorphism detection: a decomposition approach. *IEEE Transactions on Knowledge and Data Engineering*, 12(2):307–323, 2002. 31

J. Mestres. Computational chemogenomics approaches to systematic knowledge-based drug discovery. *Current Opinion in Drug Discovery & Development*, 7(3):304, 2004. ISSN 1367-6733. 6

C. Micheletti and H. Orland. MISTRAL: a tool for energy-based multiple structural alignment of proteins. *Bioinformatics*, 25(20): 2663–2669, 2009. 17, 18

K. Mizuguchi and N. Go. Comparison of spatial arrangements of secondary structural elements in proteins. *Protein Engineering, Design and Selection*, 8(4):353–362, 1995. 18

M. Moll and L.E. Kavraki. Matching of structural motifs using hashing on residue labels and geometric filtering for protein function prediction. In *CSB'08: 7th Conference on Computational Systems Bioinformatics. Proceedings*, pages 157–169, Palo Alto, USA, August 2008. 20

S.D. Mooney, M.H.P. Liang, R. DeConde, and R.B. Altman. Structural characterization of proteins using residue environments. *Proteins*, 61(4):741–747, 2005. 23

G.M. Morris, D.S. Goodsell, R.S. Halliday, R. Huey, W.E. Hart, R.K. Belew, and A.J. Olson. Automated docking using a Lamarckian genetic algorithm and an empirical binding free energy function. *Journal of Computational Chemistry*, 19(14):1639–1662, 1998. ISSN 1096-987X. 5

M.L. Moss and F.H. Rasmussen. Fluorescent substrates for the proteinases ADAM17, ADAM10, ADAM8, and ADAM12 useful for high-throughput inhibitor screening. *Analytical Biochemistry*, 366(2):144–148, 2007. 167

A. Moustafa. JAligner: open source Java implementation of Smith-Waterman. available at http://jaligner.sourceforge.net/, 2005. 107

A.G. Murzin, S.E. Brenner, T. Hubbard, and C. Chothia. SCOP: a structural classification of proteins database for the investigation of sequences and structures. *Journal of Molecular Biology*, 247(4):536–540, 1995. 15, 109, 149

R. Myers, R.C. Wilson, and E.R. Hancock. Bayesian graph edit distance. *IEEE Transactions on Pattern Analysis and Machine Intelligence*, 22(6):628–635, 2000. 33

N. Nagano, C.A. Orengo, and J.M. Thornton. One fold with many functions: the evolutionary relationships between TIM barrel families based on their sequences, structures and functions. *Journal of Molecular Biology*, 321 (5):741–765, 2002. ISSN 0022-2836. 4, 18

R. Najmanovich, N. Kurbatova, and J. Thornton. Detection of 3D atomic similarities and their use in the discrimination of small molecule protein-binding sites. *Bioinformatics*, 24(16):i105–i111, 2008. 75

R.J. Najmanovich, A. Allali-Hassani, R.J. Morris, L. Dombrovsky, P.W. Pan, M. Vedadi, A.N. Plotnikov, A. Edwards, C. Arrowsmith, and J.M. Thornton. Analysis of binding site similarity, small-molecule similarity and experimental binding profiles in the human cytosolic sulfotransferase family. *Bioinformatics*, 23(2):e104–e109, 2007. 24

H. Nakamura and S. Nishida. Numerical calculations of electrostatic potentials of protein-solvent systems by the self consistent boundary method. *Journal of the Physical Society of Japan*, 56:1609–1622, 1987. 23

T. Naumann and H. Matter. Structural classification of protein kinases using 3D molecular interaction field analysis of their ligand binding sites: target family landscapes. *Journal of Medicinal Chemistry*, 45(12):2366–2378, 2002. 6

S.B. Needleman and C.D. Wunsch. A general method applicable to the search for similarities in the amino acid sequence of two proteins. *Journal of Molecular Biology*, 48(3): 443–453, 1970. 3, 14, 82

M. Neuhaus and H. Bunke. A probabilistic approach to learning costs for graph edit distance. In *ICPR'04: 17th IAPR International Conference on Pattern Recognition. Proceedings*, volume 3, pages 389–393, Cambridge, UK, August 2004. 34

M. Neuhaus and H. Bunke. Self-organizing maps for learning the edit costs in graph matching. *IEEE Transactions on Systems, Man and Cybernetics, Part B*, 35(3):503–514, 2005. 34

M. Neuhaus and H. Bunke. A convolution edit kernel for error-tolerant graph matching. In *ICPR'06: 18th IAPR International Conference on Pattern Recognition. Proceedings*, volume 4, pages 220–223, Hong-Kong, China, August 2006a. 27, 35

M. Neuhaus and H. Bunke. A random walk kernel derived from graph edit distance. *LNCS. Structural, Syntactic, and Statistical Pattern Recognition*, 4109:191–199, 2006b. 35

C. Notredame, D.G. Higgins, and J. Heringa. T-coffee: a novel method for fast and accurate multiple sequence alignments. *Journal of Molecular Biology*, 302(1):205–217, 2000. 2, 14

R. Nussinov and H.J. Wolfson. Efficient detection of three-dimensional structural motifs in biological macromolecules by computer vision techniques. *PNAS*, 88(23):10495–10499, 1991. 16, 19

C.A. Orengo and W.R. Taylor. SSAP: sequential structure alignment program for protein structure comparison. *Methods in Enzymology*, 266:617–35, 1996. 16, 18, 117

C.A. Orengo, A.D. Michie, S. Jones, D.T. Jones, M.B. Swindells, and J.M. Thornton. CATH - a hierarchic classification of protein domain structures. *Structure*, 5(8):1093–1108, 1997. 15

C.A. Orengo, A.E. Todd, and J.M. Thornton. From protein structure to function. *Current Opinion in Structural Biology*, 9(3):374–382, 1999. 4, 18

A.R. Ortiz, C.E.M. Strauss, and O. Olmea. MAMMOTH (matching molecular models obtained from theory): an automated method for model comparison. *Protein Science*, 11 (11):2606–2621, 2002. 17, 18

P.R.J. Östergård. A new algorithm for the maximum-weight clique problem. In *TW'99: 6th Twente Workshop on Graphs and Combinatorial Optimization. Proceedings*, volume 3, pages 153–156, Enschede, The Netherlands, May 1999. 24

L. Page, S. Brin, R. Motwani, and T. Winograd. The pagerank citation ranking: Bringing order to the web. Technical report, Stanford Digital Library Technologies Project, 1998, 1998. 8

S.B. Pandit and J. Skolnick. Fr-TM-align: a new protein structural alignment method based on fragment alignments and the TM-score. *BMC Bioinformatics*, 9(1):531–542, 2008. 17, 18

A.N. Papadopoulos and Y. Manolopoulos. Structure-based Similarity Search with Graph Histograms. In *DEXA'99: Tenth International Workshop on Database and Expert Systems Applications. Proceedings*, pages 174–178, Florence, Italy, August 1999. 34

P.M. Pardalos, J. Rappe, and M.G.C. Resende. *An Exact Parallel Algorithm for the Maximum Clique Problem.* Citeseer, 1998. 31

J. Park, K. Karplus, C. Barrett, R. Hughey, D. Haussler, T. Hubbard, and C. Chothia. Sequence comparisons using multiple sequences detect three times as many remote homologues as pairwise methods. *Journal of Molecular Biology*, 284(4):1201–1210, 1998. 14

W.R. Pearson. Searching protein sequence libraries: comparison of the sensitivity and selectivity of the Smith-Waterman and FASTA algorithms. *Genomics*, 11(3):635–650, 1991. 2, 14

M. Pellecchia, D.S. Sem, and K. W
"uthrich. NMR in drug discovery. *Nature Reviews Drug Discovery*, 1(3):211–219, 2002. ISSN 1474-1776. 1

S. Pérot, O. Sperandio, M.A. Miteva, A.C. Camproux, and B.O. Villoutreix. Druggable pockets and binding site centric chemical space: a paradigm shift in drug discovery.

Drug Discovery Today, 15(15-16):656–667, 2010. ISSN 1359-6446. 2

K.P. Peters, J. Fauck, and C. Frommel. The automatic search for ligand binding sites in proteins of known three-dimensional structure using only geometric criteria. *Journal of Molecular Biology*, 256(1):201–214, 1996. 6, 21, 25, 61

B.J. Polacco and P.C. Babbitt. Automated discovery of 3D motifs for protein function annotation. *Bioinformatics*, 22(6):723–730, 2006. 18

C.T. Porter, G.J. Bartlett, and J.M. Thornton. The catalytic site atlas: a resource of catalytic sites and residues identified in enzymes using structural data. *Nucleic Acids Research*, 32 (Database issue):129–133, 2004. 19, 156

R. Potestio, T. Aleksiev, F. Pontiggia, S. Cozzini, and C. Micheletti. ALADYN: a web server for aligning proteins by matching their large-scale motion. *Nucleic Acids Research*, 38(2):W41–W45, 2010. 18

B. Qian, S. Raman, R. Das, P. Bradley, A.J. McCoy, R.J. Read, and D. Baker. High-resolution structure prediction and the crystallographic phase problem. *Nature*, 450 (7167):259–264, 2007. 2

L. Ralaivola, S.J. Swamidass, H. Saigo, and P. Baldi. Graph kernels for chemical informatics. *Neural Networks*, 18(8):1093–1110, 2005. 27, 36

A. Rangarajan and E.D. Mjolsness. A Lagrangian relaxation network for graph matching. *IEEE Transactions on Neural Networks*, 7(6):1365–1381, 2002. ISSN 1045-9227. 33

M. Rarey, B. Kramer, and T. Lengauer. Time-efficient docking of flexible ligands into active sites of proteins. In *ISMB'95: 3rd International Conference on Intelligence Systems in Molecular Biology. Proceedings*, volume 3, pages 300–308, Cambridge, UK, July 1995. 22

N.D. Rawlings, A.J. Barrett, and A. Bateman. Merops: the peptidase database. *Nucleic Acids Research*, 38(suppl 1):D227–D233, 2010. 148

J.W. Raymond and P. Willett. Maximum common subgraph isomorphism algorithms for the matching of chemical structures. *Journal of Computer-Aided Molecular Design*, 16(7): 521–533, 2002. 29

J.W. Raymond, E.J. Gardiner, and P. Willett. Heuristics for similarity searching of chemical graphs using a maximum common edge subgraph algorithm. *Journal of Chemical Information and Computer Sciences*, 42(2): 305–316, 2002. 27

R.C. Read and D.G. Corneil. The graph isomorphism disease. *Journal of Graph Theory*, 1(1):339–363, 1977. 29

I. Rechenberg and M. Eigen. *Evolutionsstrategie: Optimierung Technischer Systeme nach Prinzipien der Biologischen Evolution*. Frommann-Holzboog, 1973. 66

O.C. Redfern, A. Harrison, T. Dallman, F.M. Pearl, and C.A. Orengo. CATHEDRAL: a fast and effective algorithm to predict folds and domain boundaries from multidomain protein structures. *PLoS Computational Biology*, 3(11):2334–2347, 2007. 16, 18, 106

R.R. Regoes and S. Bonhoeffer. The HIV coreceptor switch: a population dynamical perspective. *TRENDS in Microbiology*, 13(6): 269–277, 2005. 176

A. Robles-Kelly and E.R. Hancock. Graph edit distance from spectral seriation. *IEEE Transactions on Pattern Analysis and Machine Intelligence*, 27(3):365–378, 2005. 36

B.P. Roques, M.C. Fournie-Zaluski, E. Soroca, J.M. Lecomte, B. Malfroy, C. Llorens, and J.C. Schwartz. The enkephalinase inhibitor thiorphan shows antinociceptive activity in mice. *Nature*, 288(5788):286–288, 1980. 167

B. Rost. Enzyme function less conserved than anticipated. *Journal of Molecular Biology*, 318(2):595–608, 2002. 2, 14, 18

W.P. Russ, D.M. Lowery, P. Mishra, M.B. Yaffe, and R. Ranganathan. Natural-like function in artificial WW domains. *Nature*, 437(7058): 579–583, 2005. 2, 15

R.B. Russell. Detection of protein three-dimensional side-chain patterns: new examples of convergent evolution. *Journal of Molecular Biology*, 279(5):1211–1227, 1998. 20

R.B. Russell and G.J. Barton. Multiple protein sequence alignment from tertiary structure comparison: Assignment of global and residue confidence levels. *Proteins*, 14(2): 309–323, 1992. ISSN 1097-0134. 18

R.B. Russell, P.D. Sasieni, and M.J.E. Sternberg. Supersites within superfolds. Binding site similarity in the absence of homology. *Journal of Molecular Biology*, 282(4):903–918, 1998. ISSN 0022-2836. 18

T. Ryckmans, M.P. Edwards, V.A. Horne, A.M. Correia, D.R. Owen, L.R. Thompson, I. Tran, M.F. Tutt, and T. Young. Rapid assessment of a novel series of selective CB2 agonists using parallel synthesis protocols: a lipophilic efficiency (LipE) analysis. *Bioorganic & Medicinal Chemistry Letters*, 19(15):4406–4409, 2009. ISSN 0960-894X. 6

R. Sánchez and A. Šali. Evaluation of comparative protein structure modeling by MODELLER-3. *Proteins*, 29(S1):50–58, 1997. ISSN 1097-0134. 18

O. Sander, T. Sing, I. Sommer, A.J. Low, P.K. Cheung, P.R. Harrigan, T. Lengauer, and F.S. Domingues. Structural descriptors of gp120 V3 loop for the prediction of HIV-1 coreceptor usage. *PLoS Computational Biology*, 3 (3):e58–e68, 2007. xii, 111, 175, 176, 177, 178

A. Sanfeliu and K.S. Fu. A distance measure between attributed relational graphs for pattern recognition. *IEEE Transactions on Systems, Man, and Cybernetics*, 13(3):353–362, 1983. 32, 62

J.M. Sasin, A. Godzik, and J.M. Bujnicki. SURF'S UP!—protein classification by surface comparisons. *Journal of Biosciences*, 32 (1):97–100, 2007. 25

K. Schädler and F. Wysotzki. A connectionist approach to structural similarity determination as a basis of clustering, classification and feature detection. *LNCS. Principles of Data Mining and Knowledge Discovery*, 1263/1997:254–264, 1997. 31

C. Schalon, J.S. Surgand, E. Kellenberger, and D. Rognan. A simple and fuzzy method to align and compare druggable ligand-binding sites. *Proteins*, 71(4):1755–1778, 2008. 25

E.D. Scheeff and J.L. Fink. Fundamentals of protein structure. In *Structural Bioinformatics*, chapter 2, pages 15–40. John Whiley & Sons, 2009. 4

D.C. Schmidt and L.E. Druffel. A fast backtracking algorithm to test directed graphs for isomorphism using distance matrices. *Journal of the ACM*, 23(3):433–445, 1976. 30

S. Schmitt, M. Hendlich, and G. Klebe. From structure to function: a new approach to detect functional similarity among proteins independent from sequence and fold homology. *Angewandte Chemie International Edition*, 40(17):3141–3146, 2001. 24, 106

S. Schmitt, D. Kuhn, and G. Klebe. A new method to detect related function among proteins independent of sequence and fold homology. *Journal of Molecular Biology*, 323 (2):387–406, 2002. 6, 9, 22, 24, 26, 30, 44, 50, 51, 55, 56, 59, 106, 108, 111, 117, 148

G. Schneider and U. Fechner. Computer-based de novo design of drug-like molecules. *Nature Reviews Drug Discovery*, 4(8):649–663, 2005. ISSN 1474-1776. 5

L.G. Shapiro and R.M. Haralick. Structural descriptions and inexact matching. *IEEE Transactions on Pattern Analysis and Machine Intelligence*, PAMI-3(5):504–519, 1981. 32

L.G. Shapiro and R.M. Haralick. A metric for comparing relational descriptions. *IEEE Transactions on Pattern Analysis and Machine Intelligence*, PAMI-7(1):90–94, 2009. ISSN 0162-8828. 32

D. Shasha, J.T.L. Wang, and R. Giugno. Algorithmics and applications of tree and graph searching. In *PODS'02: 21th ACM SIGMOD-SIGACT-SIGART Symposium on Principles of Database Systems. Proceedings*, pages 39–52, Madison, USA, June 2002. 27

M. Shatsky, R. Nussinov, and H.J. Wolfson. Flexible protein alignment and hinge detection. *Proteins*, 48(2):24–256, 2002a. 18

M. Shatsky, R. Nussinov, and H.J. Wolfson. MultiProt - a multiple protein structural alignment algorithm. *LNCS. Algorithms in Bioinformatics*, 2452/2002:235–250, 2002b. 17

M. Shatsky, R. Nussinov, and H.J. Wolfson. FlexProt: alignment of flexible protein structures without a predefinition of hinge regions. *Journal of Computational Biology*, 11(1):83–106, 2004. 18, 75

M. Shatsky, A. Shulman-Peleg, R. Nussinov, and H.J. Wolfson. The multiple common point set problem and Its application to molecule binding pattern detection. *Journal of Computational Biology*, 13(2):407–428, 2006. 25, 48

J. Shawe-Taylor and N. Cristianini. *Kernel Methods for Pattern Analysis*. Cambridge University Press, 2004. 34

K. Shearer, S. Venkatesh, and H. Bunke. Video sequence matching via decision tree path following. *Pattern Recognition Letters*, 22(5): 479–492, 2001. 31

Y. Shinano, T. Fujie, Y. Ikebe, and R. Hirabayashi. Solving the maximum clique problem using PUBB. In *IPPS'98: International Parallel Processing Symposium. Proceedings*, pages 326–332, Orlando, Florida, March 2002. 31

I.N. Shindyalov and P.E. Bourne. Protein structure alignment by incremental combinatorial extension (CE) of the optimal path. *Protein Engineering Design and Selection*, 11 (9):739–747, 1998. 16

I.N. Shindyalov and P.E. Bourne. A database and tools for 3-D protein structure comparison and alignment using the combinatorial extension (CE) algorithm. *Nucleic Acids Research*, 29(1):228–229, 2001. 16

A. Shulman-Peleg, R. Nussinov, and H.J. Wolfson. Recognition of functional sites in protein structures. *Journal of Molecular Biology*, 339(3):607–633, 2004. xii, 6, 24, 25, 59, 110, 136, 139, 140, 187

M. Silberstein, S. Dennis, L. Brown, T. Kortvelyesi, K. Clodfelter, and S. Vajda. Identification of substrate binding sites in enzymes by computational solvent mapping. *Journal of Molecular Biology*, 332(5):1095–1113, 2003. 22

T. Sing, N. Beerenwinkel, and T. Lengauer. Learning mixtures of localized rules by maximizing the area under the ROC curve. In *ROCAI'04: First International Workshop on ROC Analysis in Artificial Intelligence. Proceedings*, volume 22, pages 96–98, Valencia, Spain, August 2004. 176

A.P. Singh and D.L. Brutlag. Hierarchical protein structure superposition using both secondary structure and atomic representations. In *ISMB'97: 16th International Conference on Intelligence Systems for Molecular Biology. Proceedings*, volume 5, pages 284–293, Halkidiki, Greece, June 1997. 17, 18

K. Sjölander. Phylogenomic inference of protein molecular function: advances and challenges. *Bioinformatics*, 20(2):170–179, 2004. 2

T.F. Smith and M.S. Waterman. Identification of common molecular subsequences. *Journal of Bolecular Biology*, 147:195–197, 1981. 3, 14, 107

A.J. Smola and R. Kondor. Kernels and regularization on graphs. In *COLT'03: 16th Annual Conference on Learning Theory and 7th Kernel Workshop. Proceedings*, pages 144–158, Washington, DC, USA, August 2003. 35

W. Spears, K. De Jong, T. B
"ack, D. Fogel, and H. De Garis. An overview of evolutionary computation. *LNCS. Machine Learning: ECML-93*, 667: 442–459, 1993. 63

R.V. Spriggs, P.J. Artymiuk, and P. Willett. Searching for patterns of amino acids in 3D protein structures. *Journal of Chemical Information and Computer Sciences*, 43(2): 412–421, 2003. 19, 26, 30

A. Stark and R.B. Russell. Annotation in three dimensions. PINTS: patterns in non-homologous tertiary structures. *Nucleic Acids Research*, 31(13):3341–3344, 2003. 20, 57, 101

P.N. Suganthan. Attributed relational graph matching by neural-gas networks. In *NNSP'02: Neural Networks for Signal Processing X, 2000. Proceedings*, volume 1, pages 366–374, Sydney, Australia, December 2002. ISBN 0780362780. 33

P.N. Suganthan and H. Yan. Recognition of handprinted Chinese characters by constrained graph matching. *Image and Vision Computing*, 16(3):191–201, 1998. ISSN 0262-8856. 33

E. Takimoto and M.K. Warmuth. Path kernels and multiplicative updates. *The Journal of Machine Learning Research*, 4:773–818, 2003. 35

W.R. Taylor, T.P. Flores, and C.A. Orengo. Multiple protein structure alignment. *Protein Science*, 3(10):1858–1870, 1994. 16, 18

F. Teichert, U. Bastolla, and M. Porto. SABERTOOTH: protein structural alignment based on a vectorial structure representation. *BMC Bioinformatics*, 8(1):425–442, 2007. 17, 18

J.E. Thomas, C.M. Rylett, A. Carhan, N.D. Bland, R.J. Bingham, A.D. Shirras, A.J. Turner, and R.E. Isaac. Drosophila melanogaster NEP2 is a new soluble member of the neprilysin family of endopeptidases with implications for reproduction and renal function. *Biochemical Journal*, 386(2):357–366, 2005. 167

J. M. Thornton. From genome to function. *Science*, 292(5524):2095–2097, 2001. 4, 15, 18

J.M. Thornton, C.A. Orengo, A.E. Todd, and F.M.G. Pearl. Protein folds, functions and evolution. *Journal of Molecular Biology*, 293 (2):333–342, 1999. 18

J.M. Thornton, A.E. Todd, D. Milburn, N. Borkakoti, and C.A. Orengo. From structure to function: approaches and limitations. *Nature Structural & Molecular Biology*, 7: 991–994, 2000. 1

W. Tian and J. Skolnick. How well is enzyme function conserved as a function of pairwise sequence identity? *Journal of Molecular Biology*, 333(4):863–882, 2003. 2, 14

Y. Tian and J. M. Patel. TALE: a tool for approximate large graph matching. In *ICDE'08: 24th International Conference on Data Engineering. Proceedings*, pages 963–972, Cancun, Mexico, April 2008. 26, 27

Y. Tian, R.C. McEachin, C. Santos, D.J. States, and J.M. Patel. SAGA: a subgraph matching tool for biological graphs. *Bioinformatics*, 23 (2):232–239, 2007. 26, 27

A.E. Todd, C.A. Orengo, and J.M. Thornton. Evolution of function in protein superfamilies, from a structural perspective. *Journal of Molecular Biology*, 307(4):1113–1143, 2001. 2, 14, 18

C. Tonnelier, P. Jauffret, T. Hanser, and G. Kaufmann. Machine learning of generic reactions: 3. an efficient algorithm for maximal common substructure determination. *Tetrahedron Computer Methodology*, 3(6): 351–358, 1990. 30

W.H. Tsai and K.S. Fu. Error-correcting isomorphisms of attributed relational graphs for pattern analysis. *IEEE Transactions on Systems, Man and Cybernetics*, 9(12):757–768, 1979. 32

W.H. Tsai and K.S. Fu. Subgraph error-correcting isomorphism for syntactic pattern recognition. *IEEE Transactions on Systems, Man and Cybernetics*, SMC-13((1)):48–62, 1983. 32

Y.Y. Tseng and W.H. Li. Identification of protein functional surfaces by the concept of a split pocket. *Proteins*, 76(4):959–976, 2009. 22

Y.Y. Tseng, J. Dundas, and J. Liang. Predicting protein function and binding profile via matching of local evolutionary and geometric surface patterns. *Journal of Molecular Biology*, 387(2):451–464, 2009. 24

K. Tsuda, T. Kin, and K. Asai. Marginalized kernels for biological sequences. *Bioinformatics*, 18(Suppl. 1):S268–5276, 2002. 35

J.R. Ullmann. An algorithm for subgraph isomorphism. *Journal of the ACM*, 23(1):31–42, 1976. 30

S. Umeyama. An eigendecomposition approach to weighted graph matching problems. *IEEE Transactions on Pattern Analysis and Machine Intelligence*, 10(5):695–703, 1988. 36

S. Umeyama. Least-squares estimation of transformation parameters between two point patterns. *IEEE Transactions on Pattern Analysis and Machine Intelligence*, 13(4):376–380, 1991. ISSN 0162-8828. 24

H. Umezawa, T. Aoyagi, H. Morishima, M. Matsuzaki, and M. Hamada. Pepstatin, a new pepsin inhibitor produced by actinomycetes. *Journal of Antibiotics*, 23(5):259–262, 1970. 148

UniProt-Consortium. The universal protein resource (UniProt) 2009. *Nucleic Acids Research*, 37:D169–d174, 2009. 13, 162

M. Veeramalai and D. Gilbert. A novel method for comparing topological models of protein structures enhanced with ligand information. *Bioinformatics*, 24(23):2698–2705, 2008. 18

M. Veeramalai, Y. Ye, and A. Godzik. TOPS++ FATCAT: fast flexible structural alignment using constraints derived from TOPS+ Strings Model. *BMC Bioinformatics*, 9(1):358–370, 2008. 18

G. Verbitsky, R. Nussinov, and H. Wolfson. Flexible structural comparison allowing hinge-bending, swiveling motions. *Proteins*, 34(2):232–254, 1999. 75

M.L. Verdonk, J.C. Cole, P. Watson, V. Gillet, and P. Willett. Superstar: improved knowledge-based interaction fields for protein binding sites. *Journal of Molecular Biology*, 307(3):841–859, 2001. 22

M.L. Verdonk, P.N. Mortenson, R.J. Hall, M.J. Hartshorn, and C.W. Murray. Protein-ligand docking against non-native protein conformers. *Journal of Chemical Information and Modeling*, 48(11):2214–2225, 2008. ISSN 1549-9596. 109, 110

J.P. Vert. The optimal assignment kernel is not positive definite. Technical report, Centre for Computational Biology, Mines Paris Tech, 2008. 36

N. Villerbu, A.M. Gaben, G. Redeuilh, and J. Mester. Cellular effects of purvalanol A: A specific inhibitor of cyclin-dependent kinase activities. *International Journal of Cancer*, 97(6):761–769, 2002. 145

S.V.N. Vishwanathan, N. N. Schraudolph, R. Kondor, and K.M. Borgwardt. Graph kernels. *Journal of Machine Learning Research*, 11:1201–1242, 2010. 26, 36

G. Vriend and C. Sander. Detection of common three-dimensional substructures in proteins. *Proteins*, 11(1):52–58, 1991. 17

M. Wagener and J. Gasteiger. The determination of maximum common substructures by a genetic algorithm: application in synthesis design and for the structural analysis of biological activity. *Angewandte Chemie International Edition in English*, 33(11):1189–1192, 1994. 31

A.C. Wallace, N. Borkakoti, and J.M. Thornton. TESS: a geometric hashing algorithm for deriving 3D coordinate templates for searching structural databases. application to enzyme active sites. *Protein Science*, 6(11):2308–2323, 1997. 19

W.D. Wallis, P. Shoubridge, M. Kraetz, and D. Ray. Graph distances using graph union. *Pattern Recognition Letters*, 22(6-7):701–704, 2001. 29

K. Wang and R. Samudrala. FSSA: a novel method for identifying functional signatures from structural alignments. *Bioinformatics*, 21(13):2969–2977, 2005. 18

R. Wang, X. Fang, Y. Lu, and S. Wang. The pdbbind database: collection of binding affinities for protein-ligand complexes with known three-dimensional structures. *Journal of Medicinal Chemistry*, 47(12):2977–2980, 2004. 149

T. Wang and J. Zhou. EMCSS: A new method for maximal common substructure search. *Journal of Chemical Information and Computer Sciences*, 37(5):828–834, 1997. 31

X. Wang and J.T. Wang. Fast similarity search in three-dimensional structure databases. *Journal of Chemical Information and Computer Sciences*, 40(2):442–451, 2000. 32

Y.K. Wang, K.C. Fan, and J.T. Horng. Genetic-based search for error-correcting graph isomorphism. *IEEE Transactions on Systems, Man and Cybernetics, Part B*, 27(4):588–597, 1997. 33

Z. Wang, J. Eickholt, and J. Cheng. MULTICOM: a multi-level combination approach to protein structure prediction and its assessments in CASP 8. *Bioinformatics*, 26(7): 882–888, 2010. 2

Z.X. Wang. A re-estimation for the total numbers of protein folds and superfamilies. *Protein Engineering Design and Selection*, 11 (8):621–626, 1998. 2, 4, 15

P.P. Wangikar, A.V. Tendulkar, S. Ramya, D.N. Mali, and S. Sarawagi. Functional sites in protein families uncovered via an objective and automated graph theoretic approach. *Journal of Molecular Biology*, 326(3):955–978, 2003. 19

S. Wasserman and K. Faust. *Social Network Analysis: Methods and Applications*. Cambridge University Press, 1994. 8

J.D. Watson, G.J. Bartlett, and J.M. Thornton. Inferring protein function from structure. In *Structural Bioinformatics*, chapter 21, pages 515–537. John Whiley & Sons, 2009. 4, 7, 14

E.C. Webb. *Enzyme Nomenclature 1992: Recommendations of the Nomenclature Committee of the International Union of Biochemistry and Molecular Biology on the Nomenclature and Classification of Enzymes*. Academic Press, 1992. 149

M. Weisel, E. Proschak, and G. Schneider. PocketPicker: analysis of ligand binding-sites with shape descriptors. *Chemistry Central Journal*, 1(7):1–17, 2007. 22

N. Weskamp. *Efficient Algorithms for Robust Pattern Mining on Structured Objects with Applications to Structure-based Drug Design*. PhD thesis, Philipps-Universität Marburg, 2007. 10, 26, 33, 44, 45, 46, 47, 55, 59, 61, 62, 63, 65, 106, 112, 179

M. Westby, M. Lewis, J. Whitcomb, M. Youle, A.L. Pozniak, I.T. James, T.M. Jenkins, M. Perros, and E. Van Der Ryst. Emergence of CXCR4-using human immunodeficiency virus type 1 (HIV-1) variants in a minority of HIV-1-infected patients following treatment with the CCR5 antagonist maraviroc is from a pretreatment CXCR4-using virus reservoir. *Journal of Virology*, 80(10): 4909–4920, 2006. 176

J.C. Whisstock and A.M. Lesk. Prediction of protein function from protein sequence and structure. *Quarterly Reviews of Biophysics*, 36(3):307–340, 2004. 2, 14

M.L. Williams, R.C. Wilson, and E.R. Hancock. Deterministic search for relational graph matching. *Pattern Recognition*, 32(7): 1255–1271, 1999. 33

C.A. Wilson, J. Kreychman, and M. Gerstein. Assessing annotation transfer for genomics: quantifying the relations between protein sequence, structure and function through traditional and probabilistic scores. *Journal of Molecular Biology*, 297(1):233–249, 2000. 13, 18

R.C. Wilson and E.R. Hancock. Structural matching by discrete relaxation. *IEEE Transactions on Pattern Analysis and Machine Intelligence*, 19(6):634–648, 1997. 33

T.C. Wood and W.R. Pearson. Evolution of protein sequences and structures. *Journal of Molecular Biology*, 291(4):977–995, 1999. 13

I. Xenarios, L. Salwinski, X.J. Duan, P. Higney, S.M. Kim, and D. Eisenberg. DIP, the database of interacting proteins: a research tool for studying cellular networks of protein interactions. *Nucleic Acids Research*, 30(1): 303–305, 2002. 26

L. Xie and P.E. Bourne. A robust and efficient algorithm for the shape description of protein structures and its application in predicting ligand binding sites. *BMC Bioinformatics*, 8 (Suppl. 4):S9, 2007. ISSN 1471-2105. 24

L. Xie and P.E. Bourne. Detecting evolutionary relationships across existing fold space, using sequence order-independent profile–profile alignments. *PNAS*, 105(14):5441–5446, 2008. 24, 26, 30

L. Xu and I. King. A PCA approach for fast retrieval of structural patterns in attributed graphs. *IEEE Transactions on Systems, Man, and Cybernetics*, 31(5):812–817, 2001. ISSN 1083-4419. 36

L. Xu and E. Oja. Improved simulated annealing, boltzmann machine, and attributed graph matching. In *EURASIP'90: Workshop on Neural Networks. Proceedings*, pages 151–160, Sesimbra, Portugal, February 1990. 33

R.R. Yager. On ordered weighted averaging aggregation operators inmulticriteria decisionmaking. *IEEE Transactions on Systems, Man and Cybernetics*, 18(1):183–190, 1988. 74

X. Yan and J. Han. gSpan: graph-based substructure pattern mining. In *ICDM'08: IEEE International Conference on Data Mining. Proceedings*, pages 721–724, Pisa, Italy, December 2002. 29

X. Yan and J. Han. CloseGraph: mining closed frequent graph patterns. In *KDD'03: Ninth ACM SIGKDD International Conference on Knowledge Discovery and Data Mining. Proceedings*, pages 286–295, Washington, DC, USA, August 2003. 27, 29

X. Yan, P.S. Yu, and J. Han. Graph indexing: a frequent structure-based approach. In *SIGMOD'04: ACM SIGMOD International Conference on Management of Data. Proceedings*, pages 335–346, Paris, France, June 2004. 27

X. Yan, P.S. Yu, and J. Han. Substructure similarity search in graph databases. In *SIGMOD'05: ACM SIGMOD International Conference on Management of Data. Proceedings*, pages 766–777, Baltimore, USA, June 2005. 27

X. Yan, F. Zhu, J. Han, and P.S. Yu. Searching substructures with superimposed distance. In *ICDE'06: 22nd International Conference on Data Engineering. Proceedings*, volume 88, pages 88–88, Atlanta, USA, April 2006. 27

REFERENCES

Y. Ye and A. Godzik. Flexible structure alignment by chaining aligned fragment pairs allowing twists. *Bioinformatics*, 19(Suppl. 2): ii246–ii255, 2003. 18

C. Yeats, M. Maibaum, R. Marsden, M. Dibley, D. Lee, S. Addou, and C.A. Orengo. Gene3D: modelling protein structure, function and evolution. *Nucleic Acids Research*, 34(Database Issue):D281–D284, 2006. 15

T.I. Zarembinski, L.W. Hung, H.J. Mueller-Dieckmann, K.K. Kim, H. Yokota, R. Kim, and S.H. Kim. Structure-based assignment of the biochemical function of a hypothetical protein: a test case of structural genomics. *PNAS*, 95(26):15189–15193, 1998. 15, 18

S. Zhang, M. Hu, and J. Yang. Treepi: A novel graph indexing method. In *ICDE'07: 23rd IEEE International Conference on Data Engineering. Proceedings*, pages 966–975, Istanbul, Turkey, April 2007. 27

Y. Zhang and J. Skolnick. Scoring function for automated assessment of protein structure template quality. *Proteins*, 57(4):702–710, 2004. 17

F. Zhu, B.C. Han, P. Kumar, X.H. Liu, X.H. Ma, X.N. Wei, L. Huang, Y.F. Guo, L.Y. Han, C.J. Zheng, and Y. Chen. Update of TTD: therapeutic target database. *Nucleic Acids Research*, 38(Database issue):D787–D791, 2010. ISSN 0305-1048. 5

J. Zhu, H. Fan, H. Liu, and Y. Shi. Structure-based ligand design for flexible proteins: application of new F-DycoBlock. *Journal of Computer-aided Molecular Design*, 15(11): 979–996, 2001. ISSN 0920-654X. 5

H.X. Zou, X.L. Xie, U. Linne, X.D. Zheng, and S.M. Li. Simultaneous C7-and N1-prenylation of cyclo-l-Trp-l-Trp catalyzed by a prenyltransferase from Aspergillus oryzae. *Organic & Biomolecular Chemistry*, 8(13): 3037–3044, 2010. ISSN 1477-0520. 3

Acknowledgements

First of all, I would like to thank my supervisors Prof. Dr. Eyke Hüllermeier and Prof. Dr. Gerhard Klebe for giving me the opportunity to work on this topic and to prove myself on the field of bioinformatics. Hopefully, their trust was not misplaced.

Furthermore I would like to thank my colleague Thomas Fober, who co-developed some of the approaches, for the successful cooperation we had. My colleague Florian Finkernagel, for lively discussions, criticism and advice where desired and sarcasm where appreciated. And of course for proof-reading this thesis. Weiwei Cheng for being a good boy and for helping me out whenever I had questions related to machine learning. Ralph Moritz and Thibeaux Possompés, for the programming work on the original kernel methods.

In the Drug Design Group, I would like to thank Serghei Glinca, Sven Siebler and Dr. Simon Cottrell for their support on everything regarding CavBase, especially Serghei, for running some experiments for me with the CavBase approach.

Especially, I would like to thank my parents Michael and Roswitha, who to this day never lost faith in me and who taught me to rely on myself.

My friends Frank, Florian (the same as above), Thomas (Troll, not the same), Dominik and Boris, the beard-wearing monday-nerds (who meet every tuesday) for being so nerdy and for providing me with an outside view on everything, preventing me from loosing touch with reality. My friend Dennis, my last resort for any problem related to mathematics and computer science. My friends Stephan and Timo for helping me to see things in the right perspective.

And, most importantly, I would like to thank my love Nicole, for her support, love and encouragement, which kept me going through the difficult moments and for her persistence that made me go home in the evening.

MIX
Papier aus verantwortungsvollen Quellen
Paper from responsible sources
FSC® C105338

Printed by Books on Demand GmbH, Norderstedt / Germany